Artificial Rearing of Reduviid Predators for Pest Management

K. Sahayaraj • R. Balasubramanian

Artificial Rearing of Reduviid Predators for Pest Management

Springer

K. Sahayaraj
St. Xavier's College, Palayamkottai
Tirunelveli, Tamil Nadu, India

R. Balasubramanian
National Institute of Virology
Alappuzha, Kerala, India

ISBN 978-981-10-2521-1 ISBN 978-981-10-2522-8 (eBook)
DOI 10.1007/978-981-10-2522-8

Library of Congress Control Number: 2016955046

Printed on acid-free paper

This Springer imprint is published by Springer Nature
The registered company is Springer Nature Singapore Pte Ltd.
The registered company address is: 152 Beach Road, #22-06/08 Gateway East, Singapore 189721, Singapore

Preface

The Reduviidae are members of the suborder Heteroptera of the order Hemiptera. Reduviids (Insecta, Hemiptera, Reduviidae) are general, polyphagous predatory insect and cosmopolitan biocontrol agent worldwide. The family are almost all predatory, except for a minority that are blood-sucking species of importance as disease vectors. About 7,000 species have been described, making it one of the largest families in the Hemiptera. They dwelled in all plantation crops including forest and social forest. Reduviids are efficient predators on insect pests of crops, playing a significant role in keeping pest populations in check. Some important biological control agents are *Platymeris laevicollis* Distant, *Zelus renardii* Kolenati, *Rhynocoris marginatus* (Fab.), *Rhynocoris kumarii* Ambrose and Livingstone, *Rhynocoris fuscipes* (Fab.), *Pristhesancus plagipennis* Walker, *Blaptostethus pallescens* Poppius, *Acanthaspis pedestris* (Stål), *Catamiarus brevipennis* (Serville), *Ectomocoris tibialis* (Distant), etc. But, why culture reduviid predators? For biocontrol agent, does the instinctive answer—high growth and multiplication rate, short life spans, and good bioefficacy after rearing with artificial diet—make these predators ideal candidates for commercial production in agriculture since they have the potential to rapidly reach farmers very easily?

This reliance points to the need to better understanding how and why artificial diet work and how and why they fail. In more than 25 years of research in reduviid predators, I have found that artificial diet for reduviid predator rearing makes up one of the most complex unexplored areas of research and need of the hours. Hence, this book is written to help explain these complexities and dynamics. Reduviid predator rearing for bio-intensive integrative pest management (BIPM) remains a sector in its infancy, and key future challenges will likely emerge as the field evolves. As such, readers are encouraged to contact the authors with feedback on this book. Such contributions will undoubtedly assist the future development of the proposed topic.

This 180-page volume book *Artificial Rearing of Reduviid Predators* is a book about artificial insect diet. The book consists of 29 tables, 14 figures, and 9 color plates in 7 chapters. The distribution and diversity of reduviids in different crops are provided in the first chapter with citations. How the predators were reared under laboratory for augmentative release program, constrains and how to ratify the same was provided in the first chapter. The note of needs and benefits of artificial rearing was given at the end of the chapter. Compositions of ingredients, developmental protocol, storage methods, feeding arena, artificial diet feeding acts, and biocontrol efficiency after feeding the reduviids with artificial diet were highlighted in the second chapter. The importance of minerals, Cl^- ions, vitamins, and antimicrobial agents in artificial diet is discussed.

Of increased interest was Chap. 3, which expands on the biology of reduviids with artificial diet alone and/or along with laboratory host *Corcyra cephalonica* larvae and also provision of water. Later, it was assessed how the artificial diet alters the gut-microbial population, their hydrolytic enzyme production, macromolecular composition (carbohydrate, protein, and lipid contents), total body protein content, and its polypeptide profiles; gut content analyses with protein markers, DNA primers, and antigen–antibody interaction were studied to know how artificial diet modulates the physiology of the reduviids. In the last chapter of the book, artificially reared reduviid was released argumentatively in groundnut field, and its impact on pestiferous insects and other natural enemies of groundnut fields and also production of the oilseed, the groundnut, was recorded. These chapters, which discuss reduviid rearing, are an essential read for undergraduates and postgraduate students, technicians, amateur aquarists, researchers attempting to rear reduviids rearing with artificial diet, researchers attempting to improve mass multiplication and reduce production costs, and those in biocontrol industry looking to upscale reduviid mass production and supply to thc farmers.

Tirunelveli, Tamil Nadu, India K. Sahayaraj
Alappuzha, Kerala, India R. Balasubramanian

Acknowledgments

This book was made possible by the valuable advice and suggestions of many people with a variety of backgrounds and from different parts of the world. Their ideas, papers, and professional activities all played a fundamental role in shaping this publication. Among them, special thanks are extended to Dr. Patrick De Clercq, Ghent University, Belgium; Dr. Ferkovich and Dr. Jeffrey from USDA-ARS, USA; and Dr. A. Benniamin, Botanical Survey of India, Arunachal Pradesh, India, for their support and encouragement. We acknowledge and thank the Council for Scientific and Industrial Research (CSIR), Govt. of India, for the funding support. Special thanks go to the management of St. Xavier's College, Palayamkottai, for their laboratory facilities and support. Special thanks also go to the staff at the Crop Protection Research Centre, St. Xavier's College, Palayamkottai.

India

Dr. K. Sahayaraj
Dr. R. Balasubramanian

Contents

About the Authors

Dr. K. Sahayaraj is an associate professor of the Department of Zoology and the director of the Crop Protection Research Centre, St. Xavier's College, Palayamkottai, affiliated to Manonmaniam Sundaranar University, Tirunelveli. He is a recognized fellow of six national societies/academic bodies. Dr. Sahayaraj has published six books and one proceeding on *Ecofriendly Insect Pest Management*. He has more than 200 publications to his credit, including original research papers, book chapters, and popular articles in insect ecology, behavior, biology, and physiology, as well as numerous papers on the biological control efficacy of reduviids, botanicals, fungal pathogens, and bionanoparticles. He has been honored with several awards from regional (Best Researcher in Science, St. Xavier's College), national (Scientists of the Year, 2008, NESA, New Delhi; Young Achievers Award, 2010, by SADHNA, Solan; Bharat Seva Ratan Gold Medal Award, 2014, by GEPRA, New Delhi; Bharat Seven Rethan Gold Medal by GEPBRA, New Delhi), and international agencies (Hyoshi Environmentalist Award, Japan; YOUNG IOBC travel grant awardee).

He has guided more than 19 Ph.D. scholars and supervised two researchers. He has consistently worked to transfer laboratory findings to neighborhood farmers and to offer them guidance on BIPM. To this end, he has brought out five manuals/books in Tamil. Most recently, he has been engaged in publishing an international journal, the *Journal of Biopesticides*.

Dr. R. Balasubramanian received his bachelor's and postgraduate degrees in zoology from Madurai Kamaraj University, India, in 2000 and 2003, respectively. His research career began in year 2004 at the Fisheries College and Research Institute, Thoothukudi, Tamil Nadu, as a technical assistant, where he carried out research in shrimp culture with different diets. In 2005, Dr. Balasubramanian went on to pursue his Ph.D. at St. Xavier's College, Palayamkottai (Manonmaniam Sundaranar University), on the synthesis of artificial diet for mass production of hemipteran predators. Here, he learned about agricultural entomology, especially biological control, the bioecology of insects, advanced technologies in modern biological science, and the artificial rearing of predator insects. He has published a number of research papers on biocontrol, bioecology of insects, and oligidic diet preparation for reduviid mass production. In March 2009, Dr. Balasubramanian joined the Department of Medical Entomology and Zoology, National Institute of Virology, Pune, where he began conducting research on medical entomology. More recently, he has served as a scientist at a national institute in Kerala field unit, Alappuzha, Kerala.

Abbreviations

AD	Artificial diet
AMG	Anterior midgut
API	Access proportion index
AT	Approaching time
BIPM	Bio-intensive integrative pest management
CBB	Coomassie Brilliant Blue
CBR	Cost benefit ratio
CC	*Corcyra cephalonica*
CFU	Colony-forming unit
CH	*Chrotogonus* sp.
CT	Consumption time
DC	Digestible carbohydrates
DD	Degree days
DGGE	Denaturing gradient gel electrophoresis
DP	Dietary protein
ELISA	Enzyme-linked immunosorbent assay
EOD	Extraoral digestion
EV	*Earias vitella*
FPI	Food preference index
HMP	Total heterotrophic bacterial population
hrs	Hours
HT	Handling time
IAA	Isoamyl alcohol
IPM	Integrated pest management
IRS	Insect ringer's solution
LPS	Lipopolysaccharide
MAb	Monoclonal antibody
MD	Meridic diet
mg	Microgram
min	Minutes
ml	Microliter
NA	Nutrient agar
NEL	Number of eggs laid
NES	Number of egg batches

NNH	Number of nymphs hatched
NRR	Net reproductive rate
NS	Not significant
OD	Oligidic diets (AD, artificial diet)
OO	*Odontotermes obesus*
OVI	Oviposition
PCR	Polymerase chain reaction
PMG	Posterior midgut
PMv	Perimicrovillar membrane
PO	Phenol oxidase
POVI	Post-oviposition periods
PRE	Preoviposition
rc	Capacity for natural increase
RFLP	Restriction fragment length polymorphism
RH	Relative humidity
RT-PCR	Real-time polymerase chain reaction
SGC	Salivary gland complex
SL	*Spodoptera litura*
T	Generation time
TAG	Triacylglycerol
Tc	Mean length of generation
TSA	Trypticase soy agar
US$	US dollar
μm	Micrometer

Reduviid: An Important Biological Control Agent

1

Abstract

In the Indian economy, agriculture plays an important role by generating jobs and income for the country. India holds the second position in the world for agricultural production. However, in India annual losses have been reported for potential production of major crops due to insect pests. Crop losses due to these harmful organisms can be controlled by using chemical pesticides. However, the intensive use of chemical pesticides has been responsible for resistance in the insect pest population and also is a human health hazard and environmental pollution. Productivity of crops grown for human consumption is at risk due to the incidence of insect pests. Considering the importance of pest control, there has been an intense requirement for more environment-friendly and sustainable approaches using naturally occurring biocontrol agents, such as Reduviids predators, to be developed.

Reduviid predators have a wide diet that includes prey from various taxa. They consume as many insects as they are able to capture, which allows them to establish and maintain high population densities. Thus, these predators could contribute to the suppression of herbivore insects. Considering these qualities, it is a good biological control agent. For the utilization of reduviid predators in a biological control program, a large number of predators is essential. It is not feasible or possible to produce the reduviids in large number utilizing natural and laboratory host preys. Establishing effective methods to rear reduviid predator can facilitate basic and applied studies on their use in biological pest control. Through various methods and technologies, some constrains have been proposed by reduviid biologists to rear the reduviids predators. The utilization of an artificial diet is an important step toward more cost-effective rearing of predators. For the first time, an artificial diet has been proposed for rearing reduviid predators. Diets with and without insect components, oligidic, meridic, and holidic diets were considered. In this chapter, we present the distribution of reduviid predators in various agro-ecosystems. Basic information on biotic and abiotic factors of the biological traits was provided, which facilitates methods to rear the reduviids.

K. Sahayaraj, R. Balasubramanian, *Artificial Rearing of Reduviid Predators for Pest Management*, DOI 10.1007/978-981-10-2522-8_1

These all applied to the development and improvement of artificial diets for entomophagous arthropods. Government or nongovernment agencies or organic gardeners or natural gardeners may produce these insects and supply to the farmers at low cost.

Keywords
Reduviids • Distribution and diversity • Bio-control agent • Cultivable crops

1.1 Current Scenario of Crop Damage due to Insect Attack

Among the various sectors of the Indian economy, agriculture plays a prominent role by generating jobs. It recruits approximately 50 % of the entire manpower and income for the country. At present, India holds the second position in the world in agricultural production. Globally, India is one of the leading manufacturers and exporters of food, fibers, meat, and energy, and it is one of the largest producers of coffee, wheat, rice, sugarcane, oil plants, cotton, and meat. However, the agricultural sector faces systematic annual losses due to pests and diseases. Insect pests cause an estimated annual loss of 13.6 % globally, and in India the average annual loss has been estimated to be 17.5 % valued at U.S. $17.28 billion of the potential production of nine major crops (Dhaliwal et al. 2010). Despite the continuous use of chemical pesticides over a period of 60 years, insect control has not been achieved to a desirable degree. The damage caused by major inspect-pests in various crops also has been compiled and reported by Reddy and Zehr (2004). Furthermore, a number of studies have established a strong relationship between pest infestation and yield loss in various crops in India (Nair 1975; Dhaliwal and Arora 1994; Rajeswari et al. 2004; Muralidharan and Pasalu 2006; Rajeswari and Muralidharan 2006). Productivity of crops grown for human consumption is at risk due to the incidence of pests, especially pathogens and insect pests.

There has been no doubt that chemical pesticides are harmful to the human beings, animals, and the environment. Further use of chemical pesticides gives opportunities for insect pests to develop resistance and resurgence. The use of synthetic pesticides was started in 1946 with the commercial introduction of DDT (dichloro diphenyl trichloroethane) for pest control (Elzinga 1978). The ensuing decades have seen the increased use, overuse, and misuse of various types of synthetic pesticides, such as DDT, benzene hexachloride, organophosphates, and synthetic pyrethroids. As a result, the first disastrous consequence has been in relation to environmental pollution leading to contamination of soil, water, and vegetation. Persistent toxic residues have been found to contaminate soils and aquatic sediments in water bodies. More importantly, the process has resulted in a generation of toxic residues in plants and agricultural produce, such as food grains, fruits, and vegetables and even mother's milk. Human consumption of these contaminated foods constitutes an important health hazard. There have been reports of acute and chronic poisoning due to chemical pesticides in many countries (Soon 1997).

Health impairments include eye, skin, lung, cardiovascular, immunosuppressive, and neurological disorders (Rajendran 2002). In addition, there is recent evidence that "gender-warping" chemicals may lead to reproductive ailments and declining fertility.

Intensive agriculture of crops coupled with continuous use of chemical pesticides has ruled out the resistance in the insect pest population. Due to their non-specificity, chemical pesticides also eliminate other beneficial insects and natural enemies of the pest, resulting in a loss of balance in the ecosystem (Dhaliwal et al. 2010). Pesticides and insecticides are harmful to the ecological balance and should be replaced as soon as possible by natural enemies or predators. We need to think about ways to replace harmful chemical insecticides and pesticides and move towards an organic, environment-friendly method of agriculture, which will ensure and preserve the richness and fertility of the soil for the coming generations. Considering this scenario, there has been an intense requirement for more environment-friendly and sustainable approaches involving the use of naturally occurring biological control agents, such as Reduviids predators.

1.2 Reduviids in Agro-ecosystems

Certain hemipteran predators (Geocorids, Nabids, Pentatomides, etc.) feed on a rather restricted number of taxonomically related prey species, whereas others, such as reduviid predators (Insect: Hemiptera: Reduviidae), have a broad/wide diet that includes prey from various taxa, such as Hemiptera, Isoptera, Diptera, Hymenoptera, Lepidoptera, Coleoptera, Orthoptera, etc. Hence, reduviid predators are called generalist predator or general predator. One of the most important advantages is that while pest species are absent, reduviids can easily sustain in the field or migrate into the neighboring habitats by utilizing alternative sources of food (prey). Previously, Sahayaraj (2007c, 2014) reported other advantages of this predator.

Insect predators have a set of distinguished, different, and specific factors characterized by the parasitoids and also other major group of entomophagous insects. Except for predatory Hymenoptera that provision their nests with prey, predacious insects consume their prey immediately after attack. Predators occur in approximately 20 insect orders; lists vary slightly, depending on the definition of scavenging and fortuitous predation. The only insect orders that are not known to contain predatory insects are the Isoptera, Phasmatodea, Phthiraptera, Strepsiptera, and Siphonaptera (Hagen 1987).

General predators (polyphagous predators or general feeders or generalist predators) frequently are the most abundant natural enemies in agro-ecosystems. They consume invariably as many insects as they are able to capture, which allows them to establish and maintain high population densities. Thus, these predators could contribute to the suppression of herbivore insect and should be used in biological control of insect pests. Coccinellids, reduviids, mirids, anthocorids, nabids, staphylinids, carabids, dermapterans, mantides, and other lacewing families, etc., are considered generalist predators in any agro-ecosystems, such as horticulture fruits and

Fig. 1.1 Reduviid *Rhynocoris marginatus* nymph (*left*) and adult (*right*) feeding on *Phenococus solanopsis*

vegetables [tomato, bhendi, brinjal, chilies, gourds, cucumber, cowpeas, beans, peas], root and tuber vegetables [carrot, beetroot, radish, potato, tapioca], green leafy vegetables [amaranths, curry leaf, celery], Cole crops [cauliflower and cabbage], and bulb vegetables [onion of both small and large]. Reduviid predators dwell in these important agricultural crops along with social forests.

Reduviidae also are called assassin bugs, ambush bugs, or thread-legged bugs or DMK poochi, kallukatti, honeybee hunter, bee killer, caterpillar hunter, leafhopper assassin bug, etc. We listed various reduviid predators observed and recorded in field at various locations with citation. They feed on a variety of pestiferous insects from Hemiptera (Fig. 1.1), Isoptera, Lepidoptera, Coleoptera, Orthopter, Hymenoptera, Diptera, etc. (Dejean et al. 2013). For the benefit and convenience of the readers, reduviid predators inhabiting crops are listed below.

1.2.1 Horticulture Crops

Horticulture is the branch of agriculture that includes the cultivation of medicinal plant, fruits, vegetables, nuts, herbs, sprouts, mushrooms, algae, flowers, seaweeds, and non-food crops, such as grass, ornamental trees, and plants. Reduviids dwell in the horticulture plants worldwide.

1.2.2 Coconut/Palms

Platymerus rhadamanthus Gerstis are observed at coconut palms in Zanzibar and East Africa. The adults are generally found in pairs hiding near each other in the debris and axils of leaves of the crown of the coconut palm (Vanderplank 1958).

Hoyt (1962) recorded the same predator in East Africa feeding on *Oryctes monoceros* adults. *Sycanus affinis* Reut. was recorded from the coconut grove in Orissa, India (Satpathy et al. 1975). From Indian coconut farms, *Catamiarus brevipennis* (Serville) (Pawar et al. 1986); *Ectrychotes dispar* (Reuter), *Rhynocoris marginatus* (Fabricius), and *Coranus atricapillus* (Singh 1985) were recorded. In addition, *Endochusinornatus* and *Rhinocoris fuscipes* were recorded from India (Sathiamma et al. 1998), which feed on the lace bug *Stephanitis typica* (Distant).

Coranus spiniscutis (Reuter) and *Cosmocolopius nigroannulatus* Stål were observed in Brazil (Jahnke et al. 2006). Similarly, *Acacia mangium*, *Sycanus leucomesus* Walker (Sajap et al. 1999), *Cosmelestes picticeps* (Cheong et al. 2010), and *Sycanus dichotomus* (Zulkefli et al. 2004; SitiNurulhidayah and Norman 2012) were recorded in oil palm of Malaysia. Different life stages of the reduviid bugs, *Caiotes versicolar*, were recorded as predators of the weevils of oil palm in Kerala (India) (http://krishikosh.egranth.ac.in/bitstream/1/2026861/1/ICAR%2068.pdf).

1.2.3 Peas and Bean

The pigeon pea (*Cajanus cajan* L.) is a perennial legume that was domesticated in South Asia at least 3500 years ago. Its seeds have become a common food grain in Asia, Africa, and Latin America. In 1949, Bose recorded *Coranus spiniscutis* (Reuter) in Cowpea agro-ecosystem in India. In Nigeria, a large number of reduviids were observed from the cowpea agro-ecosystem (Ezueh 1991). During 2003, Claver (2003) listed names of reduviids dwelled in pigeon peas at Tamil Nadu, India. In South Africa, major arthropod natural enemy populations were recorded at cowpea from vegetative to maturity stages. Reduviid, *Rhinocoris segmentarius* constituted 18 % compared with Coccinellidae (50 %), wasps (28 %), and spiders (Arachnida 11 %) (Niba 2011). *Coranus* spp., *Rhynocoris marginatus*, *Rhynocoris fuscipes*, *Paralenaeus pyrrhomelas*, *Ectomocoris* (Claver 2011) and *Rhynocoris marginatus* (Fabricius), *Irantha armipes* (Stål), *Sycanus pyrrhomelas* (Walker), and *Rhynocoris longifrons* (Stål) dwelled in Pigeon pea ecosystems in India (Ambrose and Claver 2001). In 2014, Satyagopal and coworkers recorded various reduviids (not indicating their names) in redgrams in India.

In Ecuador, *Zelus* sp. was recorded by Valarezo et al. (2009). Later, the infested corn plants cultivated in household vegetable gardens at San Miguel de Tucuman, Argentina, during the middle of spring 2012 (Virla et al. 2015). They observed *Zelus obscuridorsis* (Stal), a poorly known species already recorded from Argentina (Wygodzinsky 1957) without a certain locality of collection. *Zelus longipes* Linn. were collected from corn fields in Campinas, SE Brazil, during February to March 1998 and January 1999 (Cogni et al. 2002). Reduviid population was frequently high in corn-bean polyculture systems (Altieri and Nicholls 2004).

1.2.4 Banana

A survey was conducted from March 2002–August 2003 at West Sumatra. It showed many predators, including spiders, lady beetles or coccinellids, lacewings, reduviids, ants, and parasitic flies. Wasps were the most important predators (Hasyim et al. 2009).

1.2.5 Bhendi

Singh and Sing (1987) observed *Rhynocoris fuscipes* (Fabricius) from India. The authors also observed this reduviid in many parts of Tamil Nadu, which mainly feed *Helicoverpa armigera* Hubner in bhendi field.

1.2.6 Cabbage

In cabbage plants, *Cosmolestis picticeps* and *Rhinocoris flavipes* were observed at field level in Malaysia (http://www.fao.org/3/a-ag117e/AG117E09.htm). Polybia sp. (Miranda Ortiz 2011) and *Rhynocoris segmentarius* (Suh Niba 2011) were recorded from Sweden and South Africa. *Sinea* spp. was recorded from cabbage ecosystems in Tisey and Nicaragua (Miranda et al. 2011). Furthermore, they reported that they were not frequently found in the field. The diamondback moth *Plutella xylostella* (L.) (Lepidoptera: Plutellidae) is a serious pest of economically important cabbage (*Brassica oleracea* Linne). Reduviids were recorded from the field (Miranda et al. 2011).

1.2.7 Groundnut

From India, *Rhynocoris longifrons* (Stal) (Sahayaraj and Raju 2003), *Ectomocoris cordiger* (Stal), *Rhynocoris marginatus* (Fabricius) (Sahayaraj 1995; Sahayaraj and Paulraj 2003), *Rhynocoris kumarii* (Ambrose and Livingstone) (Sahayaraj 1994; Sahayaraj and Sivakumar 1995; Sahayaraj and Raju 2004), *Rhynocoris squaliua* (Distant) (Singh 1985), *Rhynocoris fuscipes* (Fabricius) (Singh and Sing 1987), and *Oncocephalus annulipes* Stal (Sahayaraj and Raju 2003) were observed from groundnut agro-ecosystems.

1.2.8 Citrus Plants

Rhynocoris albopunctatus Stål was observed at Uganda (Nyiira 1970). *Zelus longipes* was reported preying upon *Diaphorina citri* Kuwayama (Hemiptera: Liviidae) on citrus in central and central-east and Florida (Michaud 2002; Hall et al. 2008) and *Murraya exotica* L. (= *M. paniculata*) in south Florida (Pena et al. 2008).

Reduviids also were reported from Thailand (Clausen 1978), which feed *Papilio demoleus* L. attack *Citrus* spp. In Australia, *Pristhesancus plagipennis* (Walker) (Grundy and Maelzer 2000a, b, c, 2002) was recorded. Previously, the same reduviid was recorded from New South Wales (James 1994). A reduviid bug, *Phonoctonous principalis* was recorded on a citrus plantation at Nairobi (Kilalo 2004).

1.2.9 Orchard Trees

In California, *Zelus* is regularly found on orange trees, where it feeds on the larvae of citrus thrips *Scirtothrips* sp. (Thysanoptera: Thripidae) (Horton 1918). Two East African Assassin Bugs, *Scipinnia repax* and *Nagusta* sp., dwelled in an orchard containing *Citrus* spp., mango, *Mangifera indica* L., *Ceiba pentandra* L., and fig, *Ficus benjamina* L. They were planted in western Kenya during January 2002 (Jackson et al. 2010). In Ontario, many species of polyphagous arthropods were recorded in apple tree plantation. Among them, the most abundant predator on apple tree foliage was the reduviid *Acholla multispinosa* L. (Hagley and Allen 1990). *Agriosphodrus dohrni* Signoret were recorded from the cherry plantations at Kyoto, Japan (Inoue 1983).

1.2.10 Chillies (*Capsicum annuum*)

Rhynocoris fuscipes (Fabricius) was recorded from India by Singh and Sing (1987) in chillies agro-ecosystem.

1.2.11 Tomato

The assassin bug, *Coranus africana* El-Sebeay, was collected from clover, tomato, eggplant, and some wild plants, as *Echinochloa colonum* and *Cyndom doctylon* in wadi El-Natroun district in the western desert of Egypt (El-Sebaey and El-Wahab 2011). In India (Tamil Nadu state), the reduviid predator population (not specified by name) was 1.97 no./plant during 2003–2004 (Amutha and Manisegaran 2006). First-time *Zelus obscuridorsis* (Hemiptera: Reduviidae) was reported as a predator of the South American tomato leafminer, *Tuta absoluta* (Lepidoptera: Gelechiidae) (Speranza et al. 2014).

1.2.12 Pecan

Apiomerus floridensis Szerlip, *Arilus cristatus* (Linnaeus), *Sinea spinipes* (Herrich-Schaeffer), and *Zelus exsanguis* Stål were recorded from the United States (Mizell and Tedders 1995) in pecan.

1.2.13 Alfalfa

Observation between 2009 and 2010 revealed that alfalfa fields (cultivar WL903) at Chile, even though they were captured in nets at spring and summer, represented less than 0.5 % of the insects sampled, suggesting that this group was very rare in the fields (Ximenez-Embun et al. 2014). In 2010, Rakhshani and co-workers reported that alfalfa (*Medicago sativa* L.) was the most widely used forage crop and *Acyrthosiphon pisum* (Harris) attacked the plant worldwide. *Nagusta goedelii* (Kolenati) recorded alfalfa in Isfahan, Iran, and feeding on pea aphid, *Acyrthosiphon pisum*. The late season reduviid predator species, *Zelus renardii*, *Zelus socius* Uhler, and *Atrachelus cinereus wygodzinski, w*ere recorded in Safford, Arizona, during 1971 (Rakickas 1971).

1.2.14 Cardamom

Small cardamom *Elettaria cardamomum* Maton is grown under forest ecosystem with diverse species of natural enemies. In India, *Sycanus indacator* Stål, *Rhynocoris longifron* Stål, *Endochus migratorius* Distant, *Endochus atric*abillus, Lanca sp., *Rihirbus trichantricus*, *Epidaus bicolor*, *Acanthaspis siva* Distant, and *Ectomocoris tibialis* were observed from the cardamom during 2010–2011 in Indian Cardamom Research Institute, Spices Board, Myladumpara, India (Nagarajan and Varadarasan 2013).

1.2.15 Cashew Tree (*Anacardium occidentale*)

Six harpactorinae reduviids, such as *Endochus albomaculatus* Stål, *Epidaus bicolor* Distant, *Euagoras plagiatus* Burmeister, *Irantha armipes*, *Panthous bimaculatus* Distant, and *Sphedanolestes signatus*, were recorded from Directorate of Cashew Research, Puttur, Karnataka State of southern India (Srikumar et al. 2014). They reported that all six reduviids were found feeding on *Helopeltis antonii* Signoret (Hemiptera: Miridae) in cashew. *Rihirbus trochantericus* Stal. var. *luteous* (Harpactorinae) is one of the common predators recorded in the cashew ecosystem (Bhat et al. 2013).

1.2.16 Cocoa

Carcinoma astrologus (Sign.) *Oncocephalus subspinosus* Aniyot Ghana Re'p.; *Phonoctonus* sp., *Polytoxus walbergi* St., *Rhynocoris bicolor* F., *Rhynocoris loratus* St. *Rhynocoris obtusus* (de Beauv.), *Rhynocoris tristis* St. *Sastrapoda vicina* Schout. *Sphedanolestes picturellus* Schout *Vestula lineaticeps* (Sign.). A group of harpactorines comprising Authenta (*Authenta fulvipennis* Horvath),

Margasus (*Margasus aftelli* Stal and *Margasus impiger* Bergroth), Nagusta (*Nagusta bunctaticollis* Stal), and Odontogonus (*Odontogonus dimensis* Schouteden) were observed in the cocoa farms at Aburi, Mampong, Arnrnonokrom, and Tafo (Louis 1974). *Ectrichodia barbicornis* (Fabricius) (Ectrichodiinae), *Microstemma atrocyanea* Signoret (Ectrichodiinae), *Acanthaspis sulcipes* Signoret, *Platymeris horrida* Stål (both Reduviinae), *Ectomacoris pretiosus* Schouteden (Piratinae), and *Phonoctonus* spp. (Harpactorinae) were observed in the shrubs of the cocoa farms.

1.2.17 Tea

Acanthaspis quinquespinosa (Fab.), *Rhynocoris marginatus* (Fab.), and *Epidaus* sp. (Das et al. 2010) were recorded at a tea plantation in India.

1.2.18 Cotton

Coranus triabeatus (Hozwath) population was recorded on the maize/cowpea and was significantly higher than on the adjacent cotton in Maharasthra, India (http://ipmworld.umn.edu/chapters/sharma.htm). Reduviids also were reported in Pakistan (Williamson 1998). In total, 131 individuals belonging to Harpactorinae (*Rhynocoris* spp., *Sphedanolestes* sp., and *Lophocephalus* sp.), Peiratinae (*Catamiarus* sp. and *Ectomocoris* sp.) and Reduviinae (*Acanthaspis* sp.) (Kalidhas and Sahayaraj 2012), *Coranus aegyptiacus* (Fab.) (Singh et al. 1987), *Coranus nodulosus* Ambrose and Sahayaraj (Sahayaraj (1991), *Rhynocoris fuscipes* (Fab.) (Singh and Sing (1987); *Rhynocoris marginatus* (Fab.) (Sahayaraj (1995), *Acanthaspis pedestris* (Stål), *Acanthaspis quinquespinosa* (Fab.), *Acanthaspis subrufa* (Distant) (Rajagopal 1984) were observed in the cotton agro-ecosystem in India. Similarly, in Australia, *Pristhesancus plagipennis* (Walker) (Pyke and Brown 1996; Grundy and Maelzer 2000b) and *Oncocephalus fuscirostris* Stål (Miles and Bull 2000; Murray 1982) *Pristhesancus papuensis* (Stål) (Martin and Brown (1984) were observed. *Zelus longipes*, *Zelus laticornis*, *Zelus ruficeps*, *Zelus armillatus*, *Atrachelus cinereus*. Crassicornis *Apiomerus apicalis* (Silvie et al. 2007), and *Phonoctonus nigrofasciatus* (Stål) were recorded in England (Evans 1962).

Zelus spp. were common in American cotton agro-ecosystems: for example, *Zelus renardii* Kolenati, *Zelus exsanguis* (Stål), *Zelus cervicalis* Stål, *Zelus socius* Uhler (Ables 1978) in America and *Zelus renardii* Kolenati Cisneros and Rosenheim (1997) in California. *Honoctonus fasciatus* (P. de B.) and *Phonoctonus subimpictus* (Sta°l) dwelled in Nigeria (Parker 1972), *Pisilus tipuliformis* (Fab.) West Africa Parker (1965), and *Zelus* sp. and *Sinea* sp. Brazil Gravena and Sterling (1983). Reduviids also were observed in Bt cotton. Two reduviids, namely *Zelus* spp. and *Sinea* spp., at very low densities were recorded in Tifton GA (Torres and Ruberson 2005).

In the United States, *Rhinocoris albopunctatus* Stål, a reduviid common in cotton plantations, was observed to be a predator of the larvae of *Heliothis armigera* (Hübner), *Earias biplaga* Walker, *Earias insulana* Boisduval, and readily accepted larvae of *Cryptophlebia leucotreta* Meyrick, and the pink bollworm, *Pectinophora gossypiella* (Saunders) (Nyiira 1970). Amaral Filho and Fagundes (1996) studied the biology and reproductive behavior of *Zelus longipes*; they collected specimens in soyabean, cotton, and corn fields.

1.2.19 Cumin (*Cuminum cyminum* L.)

Cumin is a native of the Levant and Upper Egypt and now is grown in countries, such as India, North Africa, China, and America. India is one of the largest producers and consumers of cumin seed. In 2014, Satyagopal and coworkers (2014b) recorded various reduviids (not indicating their names) in cumin in India.

1.2.20 Mustard

Initially, the *Coranus spiniscutis* (Reuter) (Bose 1949) was recorded in India. Later, *Lucerne* spp. and *Pirates ephippiger* White (Miles and Bull 2000; Murray 1982) were observed in Australia.

1.2.21 Maize

From India, *Coranus spiniscutis* (Reuter) (Jalali and Singh 2002), *Ectomocoris cordiger* (Stål) (Misra 1975), *Cydnocoris gilvus* (Burmeister), *Oncocephalus impudicus* (Reuter), *Coranus spiniscutis* (Reuter) (Bose 1949), and Ailanthus *Panthous bimaculatus* (Distant) (Varma 1989) were recorded in maize agro-ecosystems at various states.

1.2.22 Potato

In the United States and Mexico, a large number of reduviid species, particularly *Rhynocoris fuscipes* (F.) (Schaefer 1983) in potato, was recorded. *Coranus spiniscutis* (Reuter) dwelled in sweet potato in India (Bose 1949). Previously, *Rhynocoris fuscipes* (F.) was recorded from potato in India (Patalappa and Basavanna 1979; Pakistan Schaefer 1983).

1.2.23 Pumpkin

Rhynocoris fuscipes (Fabricius) (Cherian and Brahmachari 1941), *Rhynocoris lapidicola* Samuel and Joseph, and *Rhynocoris nysiiphagus* (Joseph 1959) were recorded from pumpkin agro-ecosystems in India.

1.2.24 Rice

Polytoxus fuscovittatus (Stål) (Satpathy et al. 1975), *Coranus spiniscutis* (Reuter) (Bose 1949), and *Rhynocoris fuscipes* (Fabricius) (Singh 1985; Viraktamath 2013) were observed at rice plantations in India. Later, reduviid predators belonging to four genera *viz*., Oncocephalus, Ectomocoris, Sirthenea, and Sycanus, were observed in rice ecosystems from September to December 2006 in Tamil Nadu, India (Ambrose et al. 2007). In 1987, *Polididus armatissimus* the most abundant reduviids was recorded in Tamil Nadu, India (Peter 1988).

1.2.25 Sesame (*Sesamum indicum* L.)

Sesame has numerous wild relatives in Africa and a smaller number in India. It is widely naturalized in tropical regions around the world and is cultivated for edible seeds, which grow in pods. In India, Satyagopal and coworkers (2014a, c) recorded various reduviids (not indicating their names) of sesame as well as redgrams.

1.2.26 Sesbania (*Sesbania sesban* (L.) Merrill)

Mesoplatys ochroptera Stål (Coleoptera: Chrysomelidae) is a serious pest of the tropical legume sesbania (*Sesbania sesban* (L.) Merrill) and is widely used for soil fertility improvement in southern Africa. Survey reveals that *Rhinocoris segmentarius* (Germar) is one of the predators of *Mesoplatys ochroptera* for the first time (Sileshi et al. 2001).

1.2.27 Soybean

During 1979–1980, in eastern Virginia at conventional as well as drill- and double-crop-planted soybean ecosystems, population densities of three reduviids, such as *Zelus* spp., *Sinea* spp., and *Arilus* spp. were abundant (McPherson et al. 1982). In Australia, *Pristhesancus plagipennis* (Walker) nymphal instars and adults (Grundy and Maelzer 2000b; Bishop and Blood 1977) and *Pristhesancus papuensis* (Stål) (Shepard et al. 1982) were *Coranus trabeatus* Horvath (Bishop and Blood 1977). From Colombia, *Gminatus wallengreni* Stål, *Trachylestes aspericollis* (Stål), *Sastrapada australica*, and *Zelus longipes* (Lin.) (Irwin and Shepard 1980) were recorded. In the United States, *Zelus socius* (Irwin and Shepard 1980), *Sinea diadema* (Fabricius) (Slater and Baranowsky 1978), and *Sinea spinipes* (Herrich-Schaeffer) (Irwin and Shepard 1980) were recorded. *Sinea complexa* (Caudell) (Irwin and Shepard 1980) was recorded from Brazil. From India, *Sycanus indagator* (Sta°l) (Greene 1973); *Rhynocoris fuscipes* (Fab.) (Singh and Singh 1987; Singh and Gangrade 1975), *Scadra annulipes* Reuter, and *Coranus spiniscutis* (Reuter) (Bose 1949) were observed.

1.2.28 Sugarcane

In sugarcane, *Acanthaspis quinquespinosa* (Fabricius) (Butani 1958) and *Rhynocoris marginatus* (Fab.) were observed from India (Sahayaraj 1999b). From Queensland, *Pristhesancus plagipennis* (Walker) was recorded by Illingworth (1921). *Zelus* spp. was prey on a wide range of insects in cotton, soybean, alfalfa, and fruit trees. *Zelus renardii* Kolenati was not only recorded from native range, mainland North and Central America. In this report, it invaded the tropical areas of Hawaii where it was found to feed on the invasive sugarcane leafhopper, *Perkinsiella saccharicida* Kirkaldy (Weirauch et al. 2006).

1.2.29 Sunflower

Grundy and Maelzer (2000a, b, c) were recorded *Pristhesancus plagipennis* from Australia.

1.2.30 Tobacco

The field work was performed in a tobacco *Nicotiana tabacum* L. (Solanaceae) culture at Brazil in 1999. Weekly collections of adults and fifth instar nymph began when the presence of egg masses and nymphs of *Cosmoclopius nigroannulatus* Stal were first registered in the study area by Jahnke et al. (2006). Similarly, *Agriocleptes bahianus* Wygodzinsky (Marques et al. 2006) and *Apiomerus lanipes* (Fabricius) (Marques et al. 2006) were recorded at Brazil. *Lema bilineata* (Germ.) attacks tobacco in the northern tobacco-growing areas of South Africa. A few reduviid predators were found to attack the eggs, larvae, and adults of *L. bilneata* under filed conditions.

In India, *Harpactor costalis* (Stal) (Sitaramaiah et al. 1975), *Coranus atricapillus* (Singh 1985), *Coranus spiniscutis* (Reuter) (Bose 1949), and *Rhynocoris squaliua* (Distant) (Singh 1985; Rao et al. 1981) were observed. *Harpactor fuscipes* (Fabricius) is a potential, biological control agent for *Myzus persicae* (Aphididae), *Heliothis assulta* (Noctuidae), and *Spodoptera litura* (Fabricius) (Noctuidae) (Deng et al. 2014) and were recorded in tobacco fields in China (Wu et al. 2003). *Cosmoclopius curacavencis* Cobben and Wygodzinsky of tobacco at the Agricultural

Experimental Station Cerro Azul (INTA), Leandro N. Alem, Misiones (Coscaron et al. 2002).

1.3 Other Agro-ecosystems

1.3.1 Anonymous Agriculture Fields

Coranus contraries (Reuter 1881), *Nagusta goedeli* (Stal 1859), *Reduvius personatus* (Linnaeus 1758) in Field of Agricultural College at Mashhad, Iron were recorded during the survey from 2007 to 2009 by Rahimi and coworkers in 2010.

1.3.2 Ornamental Plants

Murraya exotica (= *M. paniculata*) are grown as an ornamental hedge in Florida. *Zelus longipes* adults of both sexes were collected from the *Murraya exotica* hedge by Navarrete et al. (2014).

1.3.3 Forest

In Brazil, *Heza* sp., *Apiomerus* sp., *Arilus* sp., and *Harpactor angulosus* were observed (Pereira et al. 2012). Previously, *Salyavata variegata* Amyot (McMahan 1983) was observed from Costa Rica. Total of four species's of reduviid predators (*Euagoras plagiatus*, *Endochus* sp., *Rhenocoris fuscipes,* and *Sphedanolestus aterrimus* [*Sphenoptera aterrima*]) were recorded for the first time from the teak plantation at Kerala, India (Mohanadas 1996).

1.3.4 Kudzu

In 2012, Ruberson and coworkers observed *Zelus* sp. in kudu plantation. Four species *viz.*, *Cleptocoris lepturoides* (Wolff), *Ectomocoris cordatus* (Wolff), *Peirates unipunctatus* Livingstone and Murugan, and *Spilodermus quadrinotatus* (Fabricius) were found in agro-ecosystems. Two species of *Ectomocoris viz.*, *E. gangeticus* (Bergroth) and *E. tuberculatum* Livingstone and Murugan and *Lestomerus sanctus* (Fabricius), were found in scrub jungles, semiarid zones, and adjacent agro-ecosystems. *Coranus africana* El-Sebaey was described for the first time from Egypt. It was collected from different localities, including Kom Oshim, Sharkia, Wadi El-Natrun, Giza, and Sinai. Specimens were secured from some economic plants, including tomato, clover, cotton, as well as several wild desert plants.

The nymphs and adults of *Tegea atropicta* Stål were observed clustering round galleries of *Nasutitermes exitiosus* (Hill) in a garage in Sydney that was heavily infested and badly damaged by this termite, and some were found inside the infested timbers (Casimir 1960). Similarly in Costa Rica, a neotropical assassin bug, *Salyavata variegata* Amyot, and Serville were found on more than 40 % of the carton nests of *Nasutitermes* species in a lowland rain forest (McMahan 1983).

1.3.5 *Amaranthus* spp.

Rhynocoris bicolor Fabricius, *Rhynocoris albopilosus* Sign *Rhynocoris carmelita* Stal, *Rhynocoris rapax* Stal, *Rhynocoris tropicus* Herrich-Schaffer, *Hediocoris tibialis* Stal, *Peirates* spp., *Cosmolestes pictus* Klug, *Pisilus tipuliformis* Fabricius, *Vestula lineaticeps* Sign, *Nagusta* spp., and *Harpagocoris katangae* Schouteden were recorded from the *Amaratnus* spp. Nigeria (Oke et al. 2015).

1.3.6 *Acacia meamsii*

Isyndus rericulatus Stal. was identified from *Acacia meamsii* in Fujian Province, China (Haojiet et al. 1997)

1.4 Reasons for Consideration of Reduviids in Pest Control

Predatory bugs are commercially produced as biological control agents worldwide. Furthermore, enormous commercial predators were available for farmer's utilization. Important heteropteran predators used in augmentative biological control include: *Orius insidiosus*, *Orius laevigatus* (Anthocoridae), *Macrolophus pygmaeus*, *Nesidiocoris tenuis* (Miridae), *Geocoris* spp. (Geocoridae), *Podisus*, *Perillus*, *Arma* spp. (Pentatomidae), and *Rhynocoris* spp., *Zelus* spp. (Reduviidae). Consider the above account is worldwide, it is presumed that reduviids can be considered as good biological control agents (Sahayaraj 1999a, 2002b; Sahayaraj and Ravi 2007a; Sahayaraj et al. 2015a, b). Furthermore, the following reasons also have been considered.

- Distributed all types of agro-ecosystems, social forests, and their adjacent ecosystems
- Identified as potential and top natural enemy against many economically important key pests throughout the world for many years
- They appear to be adaptable, highly dispersive, and able to colonize new areas readily
- Withstand any climatic conditions, i.e., temperatures and relative humilities
- Many reduviid predators appear to be strongly dependent on the relative abundance of the different kinds of available prey
- Although generalized predator, they are specific to pest species and their stages
- Easily migrate from one agro-ecosystem to another to search for their preys
- Having many adaptive features to capture and kill their preys

- Amenable for mass production in small scale and large scale
- Utilized as biocontrol agents under field condition in India, Egypt, Australia, United States, etc. Susceptible to common insecticides utilized by the farmers (George 1996; Sahayaraj 2007c)
- Superior perforator than a common biocontrol agent *Menochilus sexmaculatus* (Fab.) under laboratory experimentation (Sahayaraj and Raju 2006)
- Can be integrated with synthetic pesticides and biopesticides

However, augmentation is applied on a commercial scale in relatively few agricultural systems.

1.5 Routine Methods for Laboratory Rearing

Considering the above-mentioned merits, many reduviid predators were reared under laboratory conditions using natural hosts, laboratory hosts, and factitious (i.e., unnatural) foods. Initially, reduviids were reared using fruit flies, houseflies, cockroaches, grasshoppers, beetles, stored pests, etc. Later, many natural preys were tried to rarer these bugs with various life stages of pestiferous insects. This may be either in small scale (Sahayaraj and Ravi 2007b) or large scale, and the latter has not been undertaken by any scientist, company, or organization. Furthermore, many storage pests life stages were utilized for the mass production of reduviids. Among them, larvae of *Corcyra cephalonica* have been used in many studies. The data suggests that life stages of various pests have the potential to be used as food for rearing various reduviid predators. A brief account is provided in the forth coming sections.

1.5.1 Natural Preys

The use of the natural or target prey for production of the predator, usually on a host plant, is called natural rearing system. For instance, *Rhynocoris marginatus* Fab. reared using *Spodoptera litura* Fab. (Lepidoptera: Noctuidae) larvae (Sahayaraj and Paulraj 2001). In another study, *Sycanus indagator* (Stål), a reduviid predator imported from India, reared with the greater wax moth larvae (*Galleria mellonella* (L.) and also larvae of fall armyworm (*Spodoptera frugiperda* (J. E. Smith) (Bass and Shepard 1974). A harpactorine reduviid *Sinea spinipes* (Herrich-Schaeffer) was reared in the laboratory on larval beet armyworm, *Spodoptera exigua* (Hübner) (Shurtz and McPherson 2005).

The relative importance of six different types of prey *viz*., *Chrotogonus* sp. *Musca domestica*, *Bruchus theobromae*, *Odontotermes obesus*, *Camponotus compressus*—and the mixture of above five preys in the laboratory mass rearing of *R. kumarii*, a biological control agent, was assessed and *Bruchus theobromae* was suggested as the most suitable prey for the laboratory mass rearing for the conservation and augmentation of *Rhynocoris kumarii* (Ambrose et al. 1991).

1.5.2 Laboratory Hosts

Organism unlikely to be attacked by a natural enemy in its natural habitat, but it supports development of reproducing a species that is easier and less expensive to rear, with or without plant materials, is represented as systems using factitious prey. Initially, reduviids were reared in the laboratory using their natural preys. For the utilization of reduviid predators in biological control program, a large number of predators was essential. It is not feasible or possible to produce the reduviids in a large number utilizing natural preys. Therefore, many reduviid biologists recommended and utilized many laboratory hosts as food for reduviid production. Some examples are as follows.

In 1994, James reared *Pristhesancus plagipennis* using adult *Drosophila* sp., larval *Tribolium castaneum* (Herbst), and *Tenebrio molitor* (L.). First-instar *Rhynocoris albopilosus* nymphs were provided with larvae of eri silkworm as prey began feeding 6–12 h after hatching and needed 8.7 days to complete their development, whereas second and third instar nymphs took 8.0 and 6.7 days, respectively to complete their stadium (Sahayaraj et al. 2015a). *Reduvius personatus* Linnaeus cultured larvae of the flour moth *Anagasta kiehiella* (Zeller) (Javahery 2013). Development of early nymphal instars of a predator bug *Rhynocoris albopilosus* Signoret (Harpactorinae) against first and second instar larvae of eri silkworm, *Samiacynthia ricini* Boisduval (Lepidoptera: Saturniidae), was investigated under laboratory by Sahayaraj et al. 2015a.

Amphibolus venator Klug was reared on larvae of *Plodia interpunctella* Hb., *Tribolium confusum* Duv., *Trogoderma granarium* Everts, *Lasioderma serricorni* Fab., and *Rhizopertha dominica* F. The most preferable host was found to be *R. dominica* and *L. serricorni* larvae. It is followed by *T. confusum* and *P. interpunctella*, whereas the larvae of *T. granarium* appeared to be unsuitable host for this predator. *Amphibolus venator* Klug was studied when preyed on larvae of *Tribolium confusum* Duv. (Coleoptera: Tenebrionidae) under laboratory conditions. The *Amphibolus venator* (Klug) preys *Trogoderma granarium* Everts (khapra Arr.), *Tribolium castaneum* (Hbst.), *Tenebroides mauritanicus* (L.), *Tribolium confusum* Duv., *Oryzaephilus surinamensis* (L.), and *Corcyra cephalonica* (Stainston) and total developmental period was found to vary from 41 to 165 days (Hussain and Aslam 1970).

The biology of *Sycanus affinis* Reut. was investigated in the laboratory with larvae of *Corcyra cephalonica* (Stnt.) as prey. The total life-cycle averaged 106.9 days. Adult males lived for 86 and females for 69 days. The females laid up to 807 eggs each (Satpathy et al. 1975). *Amphibolus venator* Klug was reared on larvae of *Plodia interpunctella* Hb., *Tribolium confusum* Duv., *Trogoderma granarium* Everts, *Lasioderma serricorni* Fab, and *Rhizopertha dominica* F. under optimum laboratory (Abd-Elgayed et al. 2015). The most preferable host was found to be *R. dominica* and *L. serricorni* larvae, followed by *Tribolium confusum* and *Plodia interpunctella*, whereas the larvae of *T. granarium* appeared to be an unsuitable host for this predator.

1.5.2.1 Ecophysiological Factors

Rearing space and type of substrate provided predator density, prey dentistry, prey size or stage, predator sex ratio, interspecies completion, polymorphic diversity, ecotypic diversity, starvation of the predators, etc., and were considered as governing factors for rearing reduviid predators. Along with these factors, sex ratio (Sahayaraj et al. 2003a), temperature and water (Sahayaraj et al. 2003b), prey size on the biology of *Zelus longipes* L. (Cogni et al. 2002), ecotypic diversity (Sahayaraj 2007a, b) gives due importance. Rearing substrate (Claver 2000), of a number of substrates (plant and nonplant) tested for their suitability for mass-rearing *Rhynocoris kumarii*, green plant shoots gave the shortest development and preoviposition periods, greatest fresh adult body weights, and significantly greater survival and fecundity. Dry plant litter was slightly less effective. Prey size has been considered for rearing *Zelus longipes* Linn. (Cogni et al. 2002).

1.5.2.2 Abiotic Factors

Temperature is one of the most important factors that influence the development and predation of predators. The effect of temperature on the rate of development of *Pristhesancus plagipennis* Walker was determined by rearing individuals under a range of constant temperatures (22.5–35 °C) (James 1992). He reported that the fastest rates of nymphal and egg development occurred at 30 and 32.5 °C respectively. A requirement of 845.7 DD (degree days) was estimated for development from egg to adult. Nymphal survivorship was highest between 25 and 30 °C. Previously, Ali and Watson (1978) also recorded temperature impacts on the biology of *Zelus renardii*. Studies were made at constant temperatures of 20, 25, 30, and 35 °C. The highest survival rate of the nymphs and the greatest longevity of adult *Zelus renardii* occurred at 25 °C. The developmental period of each instar decreased with each increase in temperature.

Amphibolus venator Klug does not preced its development at 20 °C. However, at 27 °C and 60 RH the nymphal development was 75 days, but it was reduced at 35 °C and 70 RH. It was reduced to 60 days while *Tribolium confusum* larvae as a host (Youssef and Abd-Elgayed 2015). In China, Deng et al. (2014) studied the effect of different temperatures (15–35 °C) on the development of *Harpactor fuscipes*. Results showed that the duration of development of each state decreased as temperature increased. The longest duration of one generation, 56 days, was observed at 15 °C; 42 days was seen at 25 °C; 35 days was seen at 35 °C; and the shortest duration, 33 days, was seen at 30 °C.

The effect of different temperatures on the survival rate of the nymphs and adults increased as the temperature increased from 15 to 30 °C. The survival rate of the fifth instar nymphs was only 33 % at 15 °C but increased to 92 % at 30 °C. Survival rates of the nymphs at different instar stages decreased when the temperature increased to 35 °C. The highest adult yield rate, 85 %, was observed at 30 °C, whereas the lowest yield, 18 %, was observed at 15 °C. The highest egg number, 122 eggs, was achieved at 30 °C. The highest hatching rate, 88 %, also was observed at 30 °C, whereas the lowest rate, 62 %, was seen at 20 °C. The longest life span of the adults was observed at 15 °C (males, 45 days; females, 51 days). Average life span of the males and females at 35 °C was 24 and 30 days, respectively, significantly lower from those at other temperatures.

1.6 Devices Used for Reduviid Mass Production

Utilizing natural, laboratory, and other hosts, many methodologies were developed to rear the reduviid predators in large number. Many devices as well as changes both in abiotic and biotic factors were made to rear the reduviids. Some special devices utilized so far for mass production were as follows:

1. Larval card methods (Lakkundi 1989; Lakkundi and Parshad 1987; Sahayaraj 1998, 2002a, b)
2. Group rearing (Sahayaraj 2002a; Sahayaraj and Ravi 2007b)
3. Altering the rearing substrates (Ambrose and Maran 1999; Ambrose 2000a, b)
4. Modulate rearing space (http://enbii.cu.edu.eg/Pests/IPM/paper_8_27.pdf)
5. Changing temperatures (Ali and Watson 1978)
6. Altering the relative humidity (http://enbii.cu.edu.eg/Pests/IPM/paper_8_27.pdf)
7. Various diet reared laboratory host *Corcyra cephalonica* (Sahayaraj and Sathyamoorthi 2002)
8. Hot water-killed larvae of *Tenebrio molitor* (L.) and *Helicoverpa armigera* (Hübner) for *Pristhesancus plagipennis* (Walker) rearing (Grundy et al. 2000) and also with live and frozen larvae of *Corcyra cephalonica* (Sahayaraj and Jeyalakshmi 2002)
9. Cold killed larvae were utilized for maintaining the reduviids in laboratory conditions
10. Alteration of prey (Sahayaraj 2001) and predator density (Sahayaraj 2002a)

In 2000, Grundy and coworkers devised a method for mass rearing *Pristhesancus plagipennis* (Walker). It is a predator of larvae and nymphs of many pestiferous insects for which natural enemies are not commercially available. This paper describes a containerized mass-rearing method for *Pristhesancus plagipennis* using hot water-killed larvae of *Tenebrio molitor* (L.) and *Helicoverpa armigera* (Hübner). In addition, the effect of *Pristhesancus plagipennis* density during nymphal rearing and adult oviposition was investigated. The rearing method minimizes *Pristhesancus plagipennis* cannibalism, avoids the need for live insect prey, and is space- and labor-efficient. Larvae of the yellow mealworm and *Tenebrio molitor*, were the most suitable prey for minimizing nymphal development time and mortality while producing insects with the highest body weight. When reared on a diet of *Tenebrio molitor*, the optimum rearing density was 20–27 nymphs per 5-l container. This rearing density minimized nymphal mortality to 16–22 %. The optimum density for oviposition was 16 adults per 5-l container, which provided the best compromise between egg production and space utilization.

1.7 Constrains in Rearing

Establishing effective methods of rearing reduviid predator can facilitate basic and applied studies on their use in biological pest control. Through various methods and technologies, preys, space provision, modulation of abiotic factors (temperature and relative humidity), the following constrains are proposed by various reduviid biologists to rear the reduviids predators:

1. Termite-like pestiferous insects affect reduviid predator
2. Rearing the laboratory hosts is tedious and laborious
3. Collection and selection of natural host
4. Handling of hunger reduviid predator
5. Time and seasons selected for rearing
6. Conventional rearing methods require host insects fed plant or artificial diet
7. Production of host insects is expensive and makes the use of beneficial insects economically unattractive.

To ratify these constrains, artificial rearing of reduviid predators has been proposed by Sahayaraj (2011).

1.8 Artificial Rearing: Needs and Benefits

The uses of inanimate artificial foods and preferably no plant materials is considered as artificial rearings systems. It is not available for many hemipteran predators. At final, during several decades, many successes were obtained in different countries around the world, mainly with polyphagous predators. Previously various insect predators, including lady beetles, anthocorid bugs, mirid bugs, and phytoseiid mites, were produced commercially with artificial rearing besides enormous predatory bugs and mites have been reported as efficient biological control agents worldwide. The artificial rearing of predatory insects started a long time ago, with the main goal to try to obtain a mean to multiply and produce parasitoids to be released in biological control strategies. It also is a powerful tool to conduct studies on biology, physiology, and behavior of entomophages, especially predatory species.

The utilization of an artificial diet may be an important step towards a more cost-effective rearing of predators (Sahayaraj et al. 2007). Whereas several artificial diets have been developed for predatory insects (as mentioned above), far fewer attempts have been made at rearing predatory mites on artificial diets. Furthermore, Grenier and De Clercq (2003) pointed out that, whereas measures of development and reproduction of an arthropod natural enemy on an unnatural or artificial diet are indicative for the value of the diet for rearing purposes, the ultimate quality parameter of an artificially reared natural enemy is predation or parasitization efficacy. Considering the importance of reduviid predators in agriculture, constrains in rearing for the first time an artificial diet has been proposed for rearing reduviid predators.

Diets with and without insect components (e.g., whole insect bodies, hemolymph…) and the Oligidic, meridic and holidic diets:

1. *Holidic*: chemically defined diets (amino acids, fatty acids, sugars, vitamins, minerals, etc.)
2. *Meridic*: holidic base with one or more unrefined or chemically unknown substances (e.g., yeast, liver extract, etc.)
3. *Oligidic*: containing only crude organic materials (e.g., meat diets)

Many arthropod natural enemies require plants for successful development and reproduction. Because reduviid predators are zoophagous predators (except very few), there would no longer be a need to maintain large surfaces of greenhouses or to purchase (often pesticide contaminated) plant materials on the market and source ingredients.

1.9 Conclusions

In this chapter, we presented the distribution of reduviid predators in various agroecosystems, such as cotton, soybean, chilies, corn, and tomato. Basic information on biotic and abiotic factors on the biological traits were provided, which facilitates to devise methods to rear the reduviids and provide various methods followed by different scientists to rear the reduviids. We also discussed the natural, laboratory, and artificial rearing of hemipteran predators with these applied to the development and improvement of artificial diets for entomophagous arthropods. Constrains faced by reduviid biologists were highlighted, which lead to the invention of artificial diets. We listed the types of artificial diet with definitions.

1.10 Future Recommendations

- Worldwide survey of reduviids in specific crops at various agriculture zones is useful to utilize these promising agents in biological control.
- Rearing these predators is difficult due to the lack of identification to suitable preys/food.
- Government or nongovernment agencies or organic gardeners or natural gardeners must take efforts to produce these insects and supply to farmers at low cost.
- To facilitate the above-mentioned proposal, governments can provide subsidy or loans to whoever is willing to establish a center or small-scale industry.
- Quality control measures should be considered before recommending predators for augmentative biological control.

References

Abd-Elgayed AA, Youssef NA (2015) Effect of some stored insect pest species on biological aspects of the predator, *Amphibolus venator* Klug (Hemiptera: Reduviidae). Ann Agric Sci 60(1):47–51

Ables JR (1978) Feeding behavior of an assassin bug, *Zelus renardii*. Ann Entomol Soc Am 71(4):476–478

Ali ASA, Watson TF (1978) Effect of temperature on development and survival of *Zelus renardii*. Environ Entomol 7(6):889–890

Altieri MA, Nicholls CI (2004) Biodiversity and pest management in agroecosystems, 2nd edn. Food Products Press® An Imprint of The Haworth Press, New York, p 253

Amaral Filho BF, Fagundes GG (1996) Desenvolvimento e reproducao de *Zelus longipes* L. (Heteroptera: Reduviidae) em laboratório. An. Soco Entorno!. Brasil, 25: 473–478

Ambrose DP (2000a) Substrata impact on mass rearing of the reduviid predator, *Rhynocoris kumarii* Ambrose and Livingstone (Heteroptera: Reduviidae). J Entomol Res 24(4):337–342

Ambrose DP (2000b) In: Schaefer CW, Panizzi AP (eds) Heteroptera of economic importance. CRC Press LLC, Boca Raton, pp 695–712

Ambrose DP, Claver MA (2001) Survey of reduviid predators in seven pigeonpea agroecosystems in Tirunelveli, Tamil Nadu, India. Int Chickpea Pigeon Pea Newslett 8:44–45

Ambrose DP, Maran SPM (1999) Substrata impact on mass rearing of the reduviid predator *Rhynocoris marginatus* (Fabricius) (Insecta: Heteroptera: Reduviidae). Pak J Biol Sci 2(4):1088–1091

Ambrose DP, Rani R, Sahayaraj M (1991) Prey influence on the laboratory mass rearing of *Rhinocoris kumarii* Ambrose and Livingstone, a potential biological control agent (Insecta: Heteroptera: Reduviidae). *Mitteilungen aus dem Museum fur Naturkunde in Berlin*. Zoologisches Mus Institut Spezielle Zoole (Berlin) 67(2):339–349

Ambrose DP, Sudha PG, Nagaraja K, Kumar AG (2007) Biodiversity of light-attracted entomofauna in a rice agroecosystem, Tirunelveli, Tamil Nadu, South India. Insect Environ 13(3):118–120

Amutha M, Manisegaran S (2006) Evaluation of IPM modules against *Helicoverpa armigera* (Hubner). Ann Plant Prot Sci 14:22–26

Anto Claver M (2003) Insect pests and predators in pigeonpea agroecosystem of Tamilnadu, with a note on a host record of reduviid bugs. Sci Cult 69(5–6):215–220

Bass JA, Shepard M (1974) Predation by *Sycanus indigator* on larvae of *Galleria mellonella* and *Spodoptera frugiperdai*. Entomol Exp Appl 17(2):143–148

Bishop AL, Blood PRB (1977) A record of beneficial arthropods and insect diseases in south-east Queensland cotton. Pests News Sum 23:384–386

Bhat PS, Srikumar KK, Raviprasad TN, Vanitha K, Rebijith KB, Asokan R (2013) Biology, behavior, functional response and molecular characterization of *Rihirbus trochantericus* Stal var. *luteous* (Hemiptera: Reduviidae: Harpactorinae) a potential predator of *Helopeltis* spp. (Hemiptera: Miridae). Entomol News 123(4):264–277

Bose M (1949) On the biology of *Coranus spiniscutis* Reuter, an assassin bug (Heteroptera: Reduviidae). Indian J Entomol 11:203–208

Butani DK (1958) Parasites and predators recorded on sugarcane pests in India. Indian J Entomol 20:270–282

Casimir M (1960) *Tegea atropicta* Stal (Hemiptera, Reduviidae), an unusual predator of termites. Proc Linnean Soc NSW 85(2):230–232

Cheong YL, Sajap AS, Hafidzi MN, Omar D, Abood F (2010) Outbreaks of Bagworms and their natural enemies in an oil palm, *Elaeis guineensis*, plantation at Hutan Melintang, Perak, Malaysia. J Entomol 7:141–151

Cherian MC, Brahmachari K (1941) Notes on three predatory hemipteran from South India. Indian J Entomol 3:115–118

Cisneros JJ, Rosenheim JA (1997) Ontogenetic change of prey preference in the generalist predator Zelus renardii and its influence on predator-predator interactions. Ecol Entomol 22:399–407

Clausen PC (1978) Introduced parasites and predators of arthropod pests and weeds. A world review. In: Agriculture handbook No 480, USDA

Claver MA (2011) Biodiversity of pegionpea insect pests and their predatory insects in five districts of North Eastern Uttar Pradesh, India. In: Proceeding of National conference on biotechnology for sustainable development. Gorakpur University, Uttar Pradesh, pp 127–132

Claver MA (2000) Substrata impact on mass rearing of the reduviid predator, *Rhynocoris kumarii* Ambrose and Livingstone (Heteroptera: Reduviidae). J Entomol Res 24(3):119–124

Cogni R, Freitas AVL, Amaral Filho BF (2002) Influence of prey size on predation success by *Zelus longipes* L. (Heteropterea, Reduviidae). J Appl Entomol 126(2-3):74–78

Coscaron ME, Melo ME, Oi-Iashi DV (2002) Description of the egg and fourth instar of *Cosmoclopius curacavencis* (Heteropterea : Reduviidae: Harpactorinae). Faceiya 18:59–61

Das S, Roy S, Mukhopadhyay A (2010) Diversity of arthropod natural enemies in the tea plantations of North Bengal with emphasis on their association with tea pests. Curr Sci 99(10):1457–1463

Dejean A, Revel M, Azémar F, Roux O (2013) Altruism during predation in an assassin bug. Naturwissenschaften 100(10):913–922

Deng H, Tian M, Chen Z, Qiu M (2014) Biological characteristics and effect of temperature on the development and reproduction of *Harpactor fuscipes* (Hemiptera: Reduviidae) reared on *Spodoptera litura* (Lepidoptera: Noctuidae) Larvae. J Entomol Res Soc 16(2):61

Dhaliwal GS, Ramesh A (1994) Trends in agricultural insect pest management. Commonwealth Publishers, New Delhi

Dhaliwal GS, Jindal V, Dhawan AK (2010) Insect pest problems and crop losses: changing trends. Indian J Ecol 37(1):1–7

El-Sebaey IIA, El-Wahab HAA (2011) Effect of different stages of reduviid predator *Corans africana* El-sebaey (Hemiptera: Heteroptera), on the population of *Bemisia tabaci* genn in tomato field at Qalubia and Bani-swif governorates. Egypt J Agric Res 89(2):413–430

Elzinga RJ (1978) Fundamentals of entomology. Prentice Hall of India Pvt Ltd., New Delhi

Evans DE (1962) The food requirement of *Phonoctonus nigrofasciatus* stål (Hemiptera: Reduviidae). Entomol Exp Appl 5:33–39

Ezueh MI (1991) Prospects for cultural and biological control of cowpea pests. Int J Trop Insect Sci 12(5–6):585–592

George PJE (1996) Impact of chosen insecticides on three non-target reduviid biocontrol agents (Insecta: Heteroptera: Reduviidae). Ph. D. Thesis. Manonmaniam Sundaranar University, Tirunelveli, Tamil Nadu, India

Gravana S, Sterling WL (1983) Natural predation of the cotton leaf worm (Lepidoptera: Noctuidae). J Econ Entomol 76:779–784

Greene GL (1973) Biological studies of a predator Sycanus indagator: I. Life history and feeding habits. Fla Entomol 56(3):255–257

Grenier S, De Clercq P (2003) Comparison of artificially versus naturally reared natural enemies and their potential for use in biological control. In: van Lenteren J (ed) Quality control and production of biological control agents: theory and testing procedures. CABI Publishing, Wallingford, pp 115–131

Grundy P, Maelzer D (2000a) Assessment of *Pristhesancus plagipennis* (Walker) (Hemiptera: Reduviidae) as an augmented biological control in cotton and soybean crops. Aust J Entomol 39:305–309. doi:10.1046/j.1440-6055.2000.00182.x

Grundy P, Maelzer D (2000b) Predation by the assassin bug *Pristhesancus plagipennis* (Walker) (Hemiptera: Reduviidae) of *Helicoverpa armigera* (Hubner) (Lepidoptera : Noctuiidae) and *Nezara viridula* (L.) (Hemiptera : Pentatomidae) in the laboratory. Aust J Entomol 39:280–282

Grundy P, Maelzer D (2000c) Assessment of *Pristhesancus plagipennis* (Walker) (Hemiptera: Reduviidae) as an augmented biological control in cotton and soybean crops. Aust J Entomol 39:305–309

Grundy PR, Maelzer DA (2002) Augmentation of the assassin bug *Pristhesancus plagipennis* (Walker) (Hemiptera: Reduviidae) as a biological control agent for *Helicoverpa* spp. in cotton. Aust J Entomol 41(2):192–196

Grundy PR, Maelzer DA, Bruce A, Hassan E (2000) A mass-rearing method for the assassin bug *Pristhesancus plagipennis* (Hemiptera: Reduviidae). Biol Control 18(3):243–250

Hagen KS (1987) Nutritional ecology of terrestrial insect predators. In: Slansky F Jr, Rodriquez JG (eds) Nutritional ecology of insects, mites, spiders, and related invertebrates. Wiley, New York, pp 533–577

Hagley EA, Allen WR (1990) The green apple aphid, *Aphis pomi* DeGeer (Homoptera: Aphididae), as prey of polyphagous arthropod predators in Ontario. Can Entomol 122(06):1221–1228

Hall DG, Hentz MG, Adair RC Jr (2008) Population ecology and phenology of *Diaphorina citri* (Hemiptera: Psyllidae) in two Florida citrus groves. Environ Entomol 37:914–924

Haojiet W, Floyd R, Farrow R, Changfu H, Chuanbit G, Changchun L, Tiansen X (1997) Insect damage on Acacia mearnsii in China. Recent Dev Acacia Planting 240

Hasyim A, Azwana A, Syafril S (2009) Evolution of natural enemies in controlling of the banana weevil borer *Cosmopolites sordidus* Germar in west Sumatra. Indones J Agric Sci 10(2):43–53

Horton JR (1918) The citrus thrips. US Dep Agric Bull 616:1–42

Hoyt CP (1962) Report on the investigations of predators and parasites of Oryctes in East Africa (1961–1962)

Hussain S, Aslam NA (1970) Some observations on a beneficial Reduviid bug: *Amphibolus venator*, Klug (Fam. Reduviidae: Hemiptera). Agric Pak 21(1):37–42

Illingworth JF (1921) The linear bug Phadnacantha australia Kirkaldy: a new pest of sugar cane in Queensland. Queensland Bureau of Sugar Experimental Station, Division of Entomology Brisbane Bulletin 14

Inoue H (1983) Nymphal cannibalism in relation to oviposition behavior of adults in the assassin bug, *Agriosphodrus dohrni* Signoret. Res Popul Ecol 25(1):189–197

Irwin ME, Shepard M (1980) Sampling predaceous Hemiptera on soybean. In: Sampling methods in soybean entomology. Springer, New York, pp 505–531

Jackson RR, Salm K, Nelson XJ (2010) Specialized prey selection behavior of two East African assassin bugs, *Scipinnia repax* and *Nagusta* sp. that prey on social jumping spiders. J Insect Sci 10:82, available online: insectscience.org/10.82

Jahnke SM, Redaelli LR, Diefenbach LMG (2006) Internal reproductive organs of *Cosmoclopius nigroannulatus* (Hemiptera: Reduviidae). Braz J Biol 66(2A):509–512

Jalali SK, Singh SP (2002) Seasonal activity of stem borers and their natural enemies on fodder maize. Entomon 27(2):137–146

James DG (1992) Effect of temperature on development and survival of *Pristhesancus plagipennis* (Hemiptera: Reduviidae). Entomophaga 37(2):259–264

James DG (1994) Prey consumption by *Pristhesancus plagipennis* Walker (Hemiptera : Reduviidae) during development. Aust J Entomol 21(2):43–47

Javahery M (2013) Natural history of *Reduvius personatus* Linnaeus (Hemiptera: Heteroptera: Reduviidae) in North America. Munis Entomol Zool 8(2):685–703

Joseph MT (1959) Biology binomics and economic importance of some reduviids collected from Delhi. Indian J Entomol 21:46–58

Kalidhas S, Sahayaraj K (2012) Survey of reduviids in cotton agro-ecosystem of Tamil Nadu. Middle-East J Sci Res 12(9):1216–1223

Kilalo DC (2004) Survey of the arthropod complex and monitoring and management of homopteran pests of citrus (*Citrus* spp.) and their natural enemies, Doctoral dissertation, University of Nairobi, PP 149

Lakkundi NH (1989) Assessment of reduviids for their predation and possibilities of their utilization in biological control. Ph. D. Thesis, IARI, New Delhi, India
Lakkundi NH, Parshad B (1987) A technique for mass multiplication of predator with sucking type of mouth parts with special reference to reduviids. J Soil Biol Ecol 7:65–69
Louis D (1974) Biology of Reduviidae of cocoa farms in Ghana. Am Midl Nat 91:68–89
Martin WR, Brown TM (1984) The action of acephate in *Pseudoplusia includens* (Leipdoptera: Noctuidae) and *Pristhesancus papuensis* (Hemiptera: Reduviidae). Entomol Exp Appl 35:3–9
Marques OM, Gil-Santana HR, Coutinho ML, da Silva D, Junior D (2006) Predatory bugs (Hemiptera: Reduviidae: Harpactorinae) in tobacco (Nicotiana tabacum L.) in the municipal district of Cruz das Almas, Bahia. Rev Bras Zoociencias 8(1):55–69
McMahan EA (1983) Adaptations, feeding preferences, and biometrics of a termite-baiting assassin bug (Hemiptera: Reduviidae). Ann Entomol Soc Am 76(3):483–486
McPherson RM, Smith JC, Allen WA (1982) Incidence of arthropod predators in different soybean cropping systems. Environ Entomol 11(3):685–689
Michaud J (2002) Biological control of Asian citrus psyllid, *Diaphorina citri* (Hemiptera: Psyllidae) in Florida: a preliminary report. Entomol News 113:216–216
Miles M, Bull T (2000) Update from the darling downs area-wide management project. Heliothis Stateline 3:8–10
Miranda Ortiz F (2011) Biological control of diamondback moth. The roles of predators, parasitoids, and insecticide. Swedish University of Agricultural Sciences, Uppsala, p 72
Miranda F, Bylund H, Gronberg L, Linda Larsson L, Bjorkman C (2011) Population density and killing capacity by predators of eggs and larvae of the diamondback moth in Nicaragua. Environ Entomol 40(2):333–341
Misra RM (1975) Notes on *Anthia sexguttata* Fabr. (Carabidae: Coleoptera) a new predator of *Pyrausta machaeralis* Walker and *Hyblea puera* Crammer. Indian Forester 101:604
Mizell Iii RF, Tedders WL (1995) A new monitoring method for detection of the stink bug complex in pecan orchards. Proc Southeastern Pecan Growers Assoc 88:36–40
Mohanadas K (1996) New records of some natural enemies of the teak defoliator, *Hyblaea puera* Cramer (Lepidoptera: Hyblaeidae) from Kerala, India. Entomon 21(3/4):251–253
Muralidharan K, Pasalu IC (2006) Assessments of crop losses in rice ecosystems due to stem borer Damage (Lepidoptera: Pyralidae). Crop Prot 25:409–417
Murray DAH (1982) Heliothis egg parasitism study on sorghum 1982. Unpublished QDPI report. QDPI, Toowoomba
Nagarajan K, Varadarasan S (2013) Reduviid fauna associated with cardamom forest ecosystem. Insect Environ 19(1):50–53
Nair MRGK (1975) Insects and mites of crops in India. Indian Council of Agricultural Research, New Delhi
Navarrete B, Carrillo D, Reyes-Martinez AY, Sanchez-Pena S, Lopez-Arroyo J, McAuslane H, Pena JE (2014) Effect of *Zelus longipes* (Hemiptera: Reduviidae) on *Diaphorina citri* (Hemiptera: Liviidae) and Its Parasitoid *Tamarixia radiata* (Hymenoptera: Eulophidae) under controlled conditions. Fla Entomol 97(4):1537–1543
Niba AS (2011) Arthropod assemblage dynamics on cowpea (*Vigna unguiculata* L. Walp) in a subtropical agro-ecosystem South Africa. Afr J Agric Res 6(4):1009–1015
Nyiira ZM (1970) The biology and behavior of *Rhinocoris albopunctatus* (Hemiptera: Reduviidae). Ann Entomol Soc Am 63(5):1224–1227
Oke OA, Odiyi CA, Ofuya TI (2015) Insects associated with underutilized crop: grain, leafy and ornamental amaranth in Ibadan, Nigeria. J Agric Ecol Res Int 2(2):145–155
Parker AH (1965) The predatory behaviour and life history of *Pisilus tipuliformis* Fabr. (Hemiptera: Reduviidae). Entomol Exp Appl 8:1–12
Parker AH (1972) The predatory and sexual behaviour of *Phonoctomus fasciatus* and *P. subimpictus* Stål (Hemiptera: Reduviidae). Bull Entomol Res 62:139–150

Patalappa G, Basavanna PC (1979) Seasonal incidence and life history of Pediobius foveolatus (Hymenoptera: Eulophidae), a parasite of *Henosepilachna vigintioctopunctata* (Fabricius) (Coleoptera: Coccinellidae). Mysore J Agric Sci 13:191–196

Pawar CS, Bhatnagar VS, Jadhav DR (1986) Heliothis species and their natural enemies, with their potential in biological control. Proc Indian Acad Sci (Anim Sci) 95:695–703

Pena JE, Duncan R, Jacas J (2008) Dynamics of mortality factors of the citrus psyllid in South Florida. Proc Fla State Hortic Soc 121:113–117

Pereira AIA, Fadini MAM, Pikart TG, Zanuncio JC, Serraõ JE (2012) New hosts and parasitism notes for the mite Leptus (Acari: Erythraeidae) in fragments of the Atlantic Forest. Brazil Braz J Biol 72(3):611–616

Peter C (1988) New records of natural enemies associated with the brown planthopper, *Nilaparvata lugens* (Stal). Curr Sci 57(19):1087–1088

Pyke BA, Brown EH (1996) The cotton pest and beneficial guide. CRDC, Woolloongabba, pp 41–43

Rajagopal D (1984) Observation on the natural enemies of *Odontotermes wallonensis* (Wassmann) (Isoptera: Termiti-dae) in South India. J Soil Biol Ecol 4:102–107

Rajendran S (2002) Encyclopedia of pest management. Marcel Dekker, New York, pp 654–656

Rajarajeswari NVL, Muralidharan K (2006) Estimates of farm yields and district production loss from blast epidemics. J Mycol Plant Pathol 36(2):115–124

Rajarajeswari NVL, Dinaker C, Muralidharan K (2004) Assessing injury to grain filling and yield losses from tungro virus epidemics. Indian J Plant Protect 32(1):73–79

Rakickas RJ (1971) Population trends exhibited by *Lygus* spp. and seven common predators in strip-cut alfalfa. Master of Science thesis, The University of Arizona, p 48

Rao RSN, Satyanarayana SVV, Soundarajan V (1981) Notes on new addition to the natural enemies of *Spodoptera litura* F. and *Myzus persicae* Sulz. on fine cored tobacco in Andhra Pradesh. Sci Cult 47:98–99

Reddy KVS, Zehr UB (2004) Novel strategies for overcoming pests and diseases in India. In: Proceedings of the 4th international crop science congress on new directions for a diverse planet, vol 25. Brisbane, pp 1–8

Sahayaraj K (1991) Bioecology, ecophysiology and ethology of chosen predatory hemipterans and their potential in biological control (Insecta: Heteroptera: Reduviidae). Ph. D. Thesis, Madurai Kamaraj University, Madurai, India

Sahayaraj K (1994) Capturing success by reduviid predators *Rhynocoris kumarii* and *Rhynocoris marginatus* on different age groups of *Spodoptera litura*, a polyphagous pest (Heteroptera : Reduviidae). J Ecobiol 6(3):221–224

Sahayaraj K (1995) Bio-efficacy and development of a reduviid predator *Rhynocoris marginatus* Fab. on *Spodoptera litura* Fab. infesting groundnut. Int Arachis Newsl 19:41–42

Sahayaraj K (1998) Laboratory rearing of predaceous bugs with special reference to reduviids (Insecta : Hemiptera : Reduviidae). Zoo's Print 13(5):17–18

Sahayaraj K (1999a) Effect of prey and their age on the feeding preference of *Rhynocoris marginatus* (Fab.). Int Arachis Newsl 19:39–41

Sahayaraj K (1999b) Field evaluation of *Rhynocoris marginatus* (Fab.) against two groundnut defoliators. Int Arachis Newsl 19:41–42

Sahayaraj K (2001) A qualitative study of food consumption, growth and fecundity of a reduviid predator in relation to prey density. Entomol Croat 5(1 and 2):19–30

Sahayaraj K (2002a) Small-scale laboratory rearing of reduviid predator *Rhinocoris marginatus* (Fab.) (Hemiptera : Reduviidae) on *Corcyra cephalonica* Stainton larvae by larval card method. J Cent Eur Agric 3(2):137–148

Sahayaraj K (2002b) Field bioefficacy of a reduviid predator *Rhynocoris marginatus* and plant products against *Aproaerema modicella* and *Spodoptera liltura* of groundnut. Indian J Entomol 64(3):292–300

Sahayaraj K (2007a) Ecotypic variation in the biology of *Acanthaspis quinquespinosa* Fabricius 1781 (Hemiptera: Reduviidae: Reduviinae) from peninsular India. Egypt J Biol 9(1):53–59

Sahayaraj K (2007b) Isolation, identification and characterization of gut flora of three reduviid predators. Asian J Microbiol Biotechnol Environ Sci 9(4):1073–1075

Sahayaraj K (2007c) Bio safety of pesticides and biopesticides. In: Pest control mechanism of Rediviids. Oxford Book Company, Narayan Niwas, Jaipur, pp 106–107

Sahayaraj K (2011) Hunter reduviids in pest management for plantation crop. In Proceeding of National Seminar on Harmful/Beneficial Insects of Agricultural Importance, Calicut, Kerala, India, p 42–51

Sahayaraj K (2014) Reduviids and their merits in biological control. In Basic and applied aspects of biopesticides. Springer, India, p 195–214

Sahayaraj K, Jeyalakshmi T (2002) Mass rearing of *Rhynocoris marginatus* Fab. on live and frozen larvae of *Corcyra cephalonica* Stainton. Entomol Croat 6(1–2):35–49

Sahayaraj K, Paulraj MG (2001) Rearing and life table of reduviid predator *Rhynocoris marginatus* Fab. (Heteroptera : Reduviidae) on *Spodoptera litura* Fab. (Lepidoptera : Noctuidae) larvae. J Appl Entomol 125(6):321–325

Sahayaraj K, Paulraj MG (2003) Insect pest and arthropods of groundnut in relation to wind velocity. Asian J Microbiol Biotechnol Environ Sci 5(2):233–235

Sahayaraj K, Raju G (2003) Pest and natural enemy complex of groundnut in Tuticorin and Tirunelveli districts of Tamil Nadu. Int Arac News Let 23:25–29

Sahayaraj K, Raju G (2004) Diversity of reduviid predators in groundnut filed of Tamil Nadu, India. J Appl Zool Res 15(2):135–140

Sahayaraj K, Ravi C (2007a) Evaluation of reduviid predators and plant products against chosen groundnut pests. Arch Phytopathol Plant Protect 40(4):281–290

Sahayaraj K, Ravi C (2007b) Small-scale mass production strategy for a reduviid predator *Rhynocoris longifrons* Stal (Heteroptera: Reduviidae). In: Guptha VK, Verma AK (eds) Perspective in animal ecology and reproduction, vol 4. Daya Publishing House, New Delhi, pp 53–81. ISBN 9788170354598

Sahayaraj K, Sathyamoorthi P (2002) Influence of different diets of *Corcyra cephalonica* Stainton on life history of *Rhynocoris marginatus* (Fab.) (Hemiptera: Reduviidae). J Cent Eur Agric 3(1):53–62

Sahayaraj K, Sivakumar K (1995) Groundnut pest and pest stage preference of a reduviid predator *Rhynocoris kumarii* Ambrose and Livingstone (Heteroptera : Reduviidae). Fresenius Environ Bull 4:263–269

Sahayaraj K, Martin P, Karthikraja S (2003a) Suitable sex ratio for the mass rearing of reduviid predator *Rhynocoris marginatus* (Fab.). J Appl Zool Res 14(1):34–37

Sahayaraj K, Martin P, Raju G (2003b) Effect of temperature and water consumption of *Rhynocoris marginatus* Fab. (Hemiptera: Reduviidae). Entomon 28(2):175–177

Sahayaraj K, Abitha Jasmine M, Selvaraj P (2003c) Side effects of selected biopesticides on reduviid predator *Rhynocoris marginatus* Fab. Entomol Croatia 7(1–2):43–50

Sahayaraj K, Venkatesh P, Balasubramanian R (2007) Feeding behaviour and biology of a reduviid predator *Rhynocoris marginatus* (Fabricius) (Heteroptera: Reduviidae) on Oligidic Diet. Hexapoda 14(1):24–30

Sahayaraj K, Natarajan S, Chelladurai K, Jayaseeli PR, Sundaram ARS (2015a) Laboratory culture of early life stages of *Rhynocoris albopilosus* (F.) (Hemiptera: Reduviidae) using early life stages of Eri silkworm (Lepidoptera: Saturniidae). Entomol Faunistique-Faunistic Entomol 68:151–157

Sahayaraj K, Kuumar V, Avery P (2015b) Functional response of *Rhynocoris kumarii* (Heteroptera: Reduviidae) on *Phenacoccus solenopsis* (Hemiptera: Pseudococcidae) in the laboratory. Eur J Entomol 112(1):69–74

Sajap AS, Kotulai JR, Kadir HA, Hussein MY (1999) Impact of prey infected by nuclear polyhedrosis virus on a predator, *Sycanus leucomesus* Walk. (Hemiptera: Reduviidae). J Appl Entomol 123:93–97

Sathiamma B, Nair KC, Soniya VP (1998) Record of the natural enemies of the lace bug *Stephanitis typica* (Distant) a pest on coconut palm. Entomon 23(4):321–324

Satpathy JM, Patnaik NC, Samalo AP (1975) Observations on the biology and habits of *Sycanus affinis* Reut. (Hemiptera: Reduviidae) and its status as a predator. J Bombay Nat Hist Soc 72(2):589–595

Satyagopal, K., Sushil, S.N. Jeyakumar, P. Shankar, G. Sharma, O.P. Boina, D.R. Sain, S.K. Lavanya, N. Sunanda, B.S. Ram Asre, K.S. Kapoor, Sanjay Arya, Subhash Kumar, C.S. Patni, Suresh D. Ekabote, K. Rajashekarappa, M. Lakshminarayana, H. Narayanaswamy, B.K. Shivanna, N. Sathyanarayana and S. Latha (2014a) AESA based IPM package for Sesame. p 30.

Satyagopal K, Sushil SN Jeyakumar, Shankar P, Sharma G, Sain OP, Boina SK, Sunanda DR, Ram Asre BS, Kapoor KS, Arya S, Kumar S, Patni CS, Gangopadhyay S, Gangadhar Naik B, Kalleshwaraswamy CM, Patnik HP, Sahu KC, Beura SK, Mohapatra SN, Bhagat, Halepyati AS, Bheemanna B, Amaresh YS, Linga Raju S, Babu R, Kumar R, Rattan GS, Bhullar MS, Singh A, Mehta PK, Raja Reddy K, Mahadkar UV, Yadava HS, Sabalpara AN, Borad PK and Pandey RN (2014b) AESA based IPM package for Cumin. p 34

Satyagopal K, Sushil SN, Jeyakumar P, Shankar G, Sharma OP, Boina DR, Sain SK, Lavanya N, Sunanda BS, Ram, Asre, Kapoor KS, Arya S, Kumar S, Patni CS, Jacob TK, Eapen SJ, Biju CN, Dhanapal K, Ravindra H, Hanumanthaswamy BC, Linga Raju S, Babu R, Sathyanarayana N, Latha S (2014c). AESA based IPM package for Redgram. p 42

Schaefer PW (1983) Natural enemies and host plants of species in the Epilachninae. (Coleoptera: Coccinellidae) – a world list. Bull Agric Exp Stn Univ Del 445

Shepard M, McWhorter RE, King EW (1982) Life history and illustrations of *Pristhesancus papuensis* (Hemiptera: Reduviidae). Can Entomol 114:1089–1092

Shurtz RA, McPherson JE (2005) Life history and laboratory rearing of *Sinea spinipes* (Heteroptera: Reduviidae) with descriptions of immature stages. Ann Entomol Soc Am 98(1):18–36

Sileshi G, Kenis M, Ogol CKPO, Sithanantham S (2001) Predators of *Mesoplatys ochroptera* in sesbania planted-fallows in eastern Zambia. BioControl 46(3):289–310

Silvie P, Aberlenc H-P, Duverger C, Bérenger JM, Cardozo R, Gómez V (2007) Harmonia axyridis no Paraguai e novos predadores identificados no cultivo do algodoeiro. In: X Simposio de controle biologico, 30 June–4 July 2007. Brasilia, Brasil, 1 p

Singh OP (1985) New record of *Rhynocoris fuscipes* Fab. as a predator of *Diciadispa armigera* (Oliver). Agric Sci Dig 5:179–180

Singh J, Arora R, Singh AS (1987) First record of predators of cotton pests in the Punjab. J Bombay Nat Hist Soc 84:456

Singh OP, Gangrade GA (1975) Parasites, predator and diseases of larvae of Diacrisia obliqua Walker (Lepidoptera: Arctiidae) on soybean. Curr Sci 44:481–482

Singh OP, Singh KJ (1987) Record of *Rhynocoris fuscipes* Fabricius as a predator of Green stink bug, Nezara viridula Linn., infesting soybean in India. J Biol Control 1:143–146

Sitaramaiah S, Joshi BG, Prasad GR, Satyanarayana SVV (1975) *Harpactor costalis* Stal. (Reduviidae; Heteroptera)-a new predator of the tobacco caterpillar *Spodoptera litura* (F.). Sci Cult 41(11):545–546

SitiNurulhidayah A, Norman K (2012) Influence of density on the predation activity of *Sycanusdichotomus* (Hemiptera: Reduviidae) on Mealworm *Tenebriomolitor* (Coleoptera: Tenebrionidae) in cage studies. UMT 11th International Annual Symposium on Sustainability Science and Management 09th – 11th Jul 2012, Terengganu, Malaysia, p 162–165

Slater JA, Baranowski RM (1978) How to know the true bugs (Hemiptera-Heteroptera). William C. Brown Co., Dubuque

Soon LG (1997) Integrated pest management in developing Asia. In: Dhaliwal GS, Randhawa NS, Arora R, Dhawan AK (eds) Proc Intl Conf Ecol Agri: towards Sust Devt. Chandigarh, India, pp 3–16

Speranza S, Melo MC, Luna MG, Virla EG (2014) First record of *Zelus obscuridorsis* (Hemiptera: Reduviidae) as a predator of the South American tomato Leafminer, *Tuta absoluta* (Lepidoptera: Gelechiidae). Fla Entomol 97(1):295–297

Srikumar KK, Bhat PS, Raviprasad TN, Vanitha K, Saroj PL, Ambrose DP (2014) Biology and behavior of six species of reduviids (Hemiptera: Reduviidae: Harpactorinae) in a cashew ecosystem 1. J Agric Urban Entomol 30(1):65–81

Torres JB, Ruberson JR (2005) Canopy-and ground-dwelling predatory arthropods in commercial Bt and non-Bt cotton fields: patterns and mechanisms. Environ Entomol 34(5):1242–1256

Valarezo O, Canarte E, Navarrete B, Intriago ML.(2009) Chicharrita Dalbulus maidisy su manejo en el cutivo de maiz. Plegable Divulgativo 305. INIAP. Estacion Experimental Portoviejo. Ecuador, 2009, p 6. Availble online: http://www.dicyt.com/noticeias/technologia-para-combatir-una-plaga-del-maiz-en-el-litoral-ecuatoriano

Vanderplank FL (1958) The assassin bug *Platymerus rhadamanthus* Gerst (Hemiptera: Reduviidae) a useful predator of the rhinoceros beetles Oryctes boas (F.) and *Oryctes monoceros* (Oliv.). (Coleoptera: Scarabaeidae). J Entomol Soc S Afr 21(2):309–314

Viraktamath BC (2013) In: Bhadana VP, Senguttuvelu P, Sheshu Madhav M, Mahender Kumar R, Brajendra P, Voleti SR, Jhansi Lakshmi V, Prasad MS, Muthuraman P, Prasad Babu MBB, Bentur JS, Neeraja CN (eds) Annual report 2012–13, directorate of rice research. p 118

Virla EG, Melo CM, Speranza S (2015) Preliminary Observations on *Zelus obscuridorsis* (Stal) (Hemiptera: Reduviidae) as Predator of the Corn Leafhopper (Hemiptera: Cicadellidae) in Argentina. Insects 6(2):508–513

Williamson S (1998) Understanding natural enemies; a review of training and information in the practical use of biological control. Biocontrol News Inf 19:117–125

Weirauch C (2006) Observations on the sticky trap predator *Zelus luridus* Stål (Heteroptera, Reduviidae, Harpactori-nae), with the description of a novel gland associated with the female genitalia. Denisia 19:1169–1180

Wu JW, Cai WZ, Hou TQ (2003) China tobacco insect species and integrated pest management, vol 129. China's Agricultural Science and Technology Press, Beijing

Wygodzinsky P (1957) On the Reduviidae mentioned or described by Carlos Berg (Hemiptera : Insecta). Rev Brasil de Biol 17(2):267–274

Ximenez-Embun MG, Zaviezo T, Grez A (2014) Seasonal, spatial and diel partitioning of *Acyrthosiphon pisum* (Hemiptera: Aphididae) predators and predation in alfalfa fields. Biol Control 69:1–7

Youssef NA, Abd-Elgayed AA (2015) Biological parameters of the predator, *Amphibolus venator* Klug (Hemiptera: Reduviidae) preying on larvae of *Tribolium confusum* Duv. (Coleoptera: Tenebrionidae). Ann Agric Sci (in press)

Zulkefli M, Norman K, Basri MW (2004) Life cycle of *Sycanus dichotomus* (Hemiptera: Pentatomidae)-a common predator of bagworm in oil palm. J Oil Palm Res 16(2):50–56

Feeding Behaviour of Reduviid Predators Against Artificial Diet

2

Abstract

Rhynocoris marginatus has been reported as a potential predator on various economically important agricultural pests. The predator can be used as potential biocontrol agent. Rearing and mass multiplication of the reduviids in the laboratory are an important requirement for the successful biological control program. A required number of bioagents can be obtained by rearing them either on their natural host. However, rearing of this predator on natural host is entirely impracticable owing to nonavailability of host throughout the year. Preparation of oligidic diet is only an option for mass multiplication. With trial and error, several diets were prepared and tested for the feeding preference on oligidic diet using different objects like cotton, capsules, cavity slides, and foam for providing oligidic diet for different nymphal and adult stages of *Rhynocoris marginatus* to find suitable which among them is suitable. The results revealed that cotton has been more preferred by *Rhynocoris marginatus*; hence, cotton is chosen as a source material for providing oligidic diet. Feeding preference aspects like diet preference, searching behavior, sucking time, and weight gain have been considered in the laboratory using oligidic diet before continuous rearing of reduviid predators to ensure the potential of prepared oligidic diet.

Stage preference study of *Rhynocoris marginatus* of different life stages fed *Dysdercus cingulatus*, *Spodoptera litura*, and *Corcyra cephalonica* separately was carried out with experiment of choice. Artificial diet-reared *Rhynocoris marginatus* maximum predatory rate was observed for *Spodoptera litura* adult female, and it was very low in third instar *Dysdercus cingulatus*. In this experiment *Rhynocoris marginatus* nymphs and adult have the capacity to consume more number of *Spodoptera litura* larvae, when the predator was reared on oligidic diet. Similarly insect hosts were reared by *Rhynocoris marginatus* second and fifth nymphal instars and adult consumed by more number of *Spodoptera litura* and other life stages consumed *Dysdercus cingulatus* second instar. Hence this reduviid can be used as a biological control agent in crop where *Spodoptera litura* and *Dysdercus cingulatus* are present.

K. Sahayaraj, R. Balasubramanian, *Artificial Rearing of Reduviid Predators for Pest Management*, DOI 10.1007/978-981-10-2522-8_2

Keywords
Rhynocoris marginatus • Meat-based artificial diet • Diet ingredients • Consistency • Object preference • Choice and non-choice test

2.1 Background Information

Most of the reduviid predators are carnivore except for few as reported by Stoner and co-workers (1975) which was evident by the presence of particles larger than 1 μm or lipid micelles with >15 μm diameters (Cohen 1998). It was also proclaimed that info-chemicals of preys like methyl esters, methyl palmitate, methyl linoleate, methyl oleate, and methyl stearate (fatty acid) exhibited characteristics like kairomonal probing behavior of *Peregrinator biannulipes* Montrouzier and Signoret toward the lure (Tebayashi et al. 2003). The chemical (tridecane, octacosane, 1-iododecane, octadecane, eicosane, pentacosane, heptacosane, and dotriacontane) cues of *Helicoverpa armigera* (Hubner) (Lepidoptera: Noctuidae), *Spodoptera litura* (F.) (Lepidoptera: Noctuidae), and *Mylabris pustulata* (Thunberg) (Coleoptera: Meloidae) elicited a quicker approaching behavior of the predator *Rhynocoris marginatus* than the control (Sahayaraj 2008). Similar response was also observed in *Rhynocoris fuscipes* (Nagarajan and Ambrose, 2013).

A first step Taylor and Schmidt (1996) fed first-instar spined assassin bugs *Sinea diadema* (Fab.) with water or glucose solutions which significantly delayed the onset of conspecific predation. Similarly it was also observed that some species seem to complement their diet by drinking floral (Narbona and Dirzo 2010) and extrafloral nectar (Ralston 1977; Tallamy et al. 2004) or sucking the sap from the plant by perforating plant tissue (Haviland 1931; Stoner et al. 1975). Ralston (1977), Tallamy et al. (2004), and Guillermo-Ferreira et al. (2012) briefly mentioned observations of a reduviid, *Atopozelus pallidens* drinking nectar from extrafloral nectaries (EFNs). In addition to the components of the artificial diets, the properties and state (liquid or solid) can strongly affect feeding efficiency. These observations reveal lots of scope for preparing artificial diet for this group of predators. However, the optimal conditions for mass rearing of reduviid insect are still unclear, thus limiting its application. Before rearing these bugs with meridic diet, it is imperative to study the feeding activity. The aim of the current chapter is to explain the feeding acts of reduviids and better understand the regulation of rearing conditions, preferential oviposition substrates, and optimal diet combinations vital for the successful artificial massive rearing of *Rhynocoris marginatus*. In this chapter, we discussed about foraging behavior on different insect pests and artificial diet focusing on *Rhynocoris marginatus* predator.

For other bibliographic references, see Grenier et al. (1994), Thompson and Hagen (1999), and Cohen (2004). For the rearing of the stinkbugs, *Podisus* spp., cylindrically shaped "artificial larvae," 2–4 cm long, and 0.3 cm diameter, were produced by bringing thawed or fresh diet to a stretched Parafilm M sheet and wrapping a single layer of the Parafilm around the meat-based diet paste (De Clercq and

Degheele 1992). For *Orius laevigatus*, the diets were encapsulated in Parafilm M sheet using an encapsulation device (ARS, Gainesville, USA) forming small hemispherical domes (35 μl) sealed with transparent tape. The Parafilm was stretched before encapsulation to facilitate stylet penetration by early instars of the predator (Bonte and De Clercq 2008, 2010). The same presentation with two artificial diets was successfully used for the development and reproduction of *Macrolophus caliginosus* (Vandekerkhove et al. 2006). Glycerol and starch are texture and consistency agents used in diets for predators such as coccinellids or chrysopids. In 2004, Arijs and co-workers developed different meat- and liver-based artificial diets which were designed for the generalist predator *Orius laevigatus* (Fieber) (Anthocoridae), and their suitability was compared to that of its factitious food, *Ephestia kuehniella* Zeller (Pyralidae) eggs.

2.2 Importance of *Rhynocoris marginatus* in Biological Control

The impact of predaceous arthropods in natural communities and agricultural crops has received more attention in recent years. Reduviids are very common in crops. Reduviids are polyphagous predators and are very effective biological control agents in crops and feeding on wide range of insect preys (Ambrose 1999; Sahayaraj 2000; Sahayaraj et al. 2002, 2006; Irudayaraj et al. 2003). *Rhynocoris marginatus* have the following characteristics:

1. Adaptability to the varying physical conditions of the environment
2. Its searching capacity, including the general mobility
3. Power of increase relative to that of its prey (host)
4. Rate of prey consumption and also the ability to survive host-free periods

Since it is a polyphagous predator, it can survive in disturbed situations such as crops where the prey species is frequently unavailable by many factors; a highly prey-specific predator will suffer a lot, whereas a more generalist-type predator like reduviids will maintain itself on other species and may respond to an available pest species when it starts to increase again. Predators are very important in biological control and in natural control. This limited attention has been given to the actual details of searching and prey selection behavior of predators (Sahayaraj and Balasubramanian 2008). *Rhynocoris marginatus* developed resistance against pesticides and biopesticides (Sahayaraj and Paulraj 1999a, b; Sahayaraj and Karthikraja 2003; Sahayaraj et al. 2002, 2003; 2012; Allaluya Jasmine et al. 2012) in Tamil Nadu, India. It is amenable for the mass production, using various preys (Ambrose et al. 1990; Sahayaraj and Paulraj 2001b), techniques, and different environmental conditions (Sahayaraj and Paulraj 2001a). In 2012, Sahayaraj wrote a review on "Possible utilization of *Rhynocoris marginatus* (Fab.) in crops pest management" where desirable and non-desirable characters are spelled out very clearly.

There are survival factors that determine prey suitability for insect predators, which can be divided into nutritional and non-nutritional factors (Sahayaraj et al.

2004). Under laboratory conditions, *Rhynocoris marginatus* was noted to have a preference for slow-moving soft-bodied insect preys and high-nutrient preys, particularly foliage-feeding lepidopteran larvae (Ambrose 2006; Sahayaraj et al. 2004; Sahayaraj and Balasubramanian 2008). The suitability of a certain prey type may not be related to nutritional quality or acceptability for the predator but to the prey-predator size ratio (Hagler and Naranjo 2005).

However, generalized predators usually show some degree of nutritional stringency and may display distinct food preference in choice situations (De Clercq 2004). The polyphagous predator *Rhynocoris marginatus* demonstrates a high level of nutritional plasticity and has the capability to utilize artificial foods, although this may result in some negative effect on biological parameters as compared with feeding on optimal insect's foods, like larvae of *Corcyra cephalonica* and *Spodoptera litura*. Previously no artificial diet has been reported to support complete development of this predator.

2.2.1 Feeding Behavior Against Meridic Diet

The reduviid predators are an important general predator that is widely used for the biological control of insect pests. However, a key limitation to the extensive use of *Rhynocoris marginatus* as a biocontrol agent, particularly of pests of food crops, is being cost effective, laborious, and time consuming for mass rearing. Between the two tested lepidopteran preys, *Corcyra cephalonica* size was smaller than the *Spodoptera litura*. Moreover, *Dysdercus cingulatus* are characterized by a heavily sclerotized integument and less preferred by predator *Rhynocoris marginatus*. Moreover, it moves very faster than lepidopteran caterpillars.

The superior fecundity rates of predators fed with *Corcyra cephalonica* suggest that it is a suitable food for the laboratory production of *Rhynocoris marginatus*. Furthermore, it was also suggested by Sahayaraj et al. (2004) that *Helicoverpa armigera* could be used for the laboratory rearing of this bug. However, the cost of mass production of *Rhynocoris marginatus* either with *Corcyra cephalonica* or *Helicoverpa armigera* was higher. Furthermore, rearing of either natural or factitious host is a laborious, tedious, and time-consuming one. Better understanding of the requirements for rearing of stable laboratory populations of *Rhynocoris marginatus* is required for the mass production of these insects for inundative release. In view of these circumstance, alternative food is required to overcome these problems that artificial diet might be a right choice for mass rearing and inundative release of this predator in the field.

2.3 Meridic Diet (MD) Development

The conventional methods and rearing diets of insects suffer from various drawbacks are discussed below:

1. The larval card method cannot be utilized for mass multiplication of reduviids on a commercial scale.
2. Success of larval card method purely depends upon the perfection of card preparation. If the card is not rightly prepared, the *Corcyra* larvae may change into pupa which is not the food for reduviid predators, or it may lead to larval escape, leading to poor feeding accessibility and, subsequently, death of reduviid predators.
3. These conventional methods require collection of prey insects from various agro-ecosystems after spending time and energy in order to feed the reduviid predators.
4. These conventional methods are unable to utilize those prey insects as a feed which have any offensive and/or defensive mechanism.
5. These conventional methods may require pest species to be reared on their natural host plants for uninterrupted supply of food to them.

In view of the above, there is a need to develop an artificial diet for rearing of reduviids which is economical, nutritious, and effective in mass production of reduviids with predatory efficacy.

The meridic diet (MD) can be prepared at low cost using meat and pig liver as the main food, along with several chemical components incorporated in the rearing medium. We formulated an oligidic diet that contains meat-based diet. The meat liver, blood, and serum were used as source ingredients. The meat-based diet tested in this study was able to sustain the *Rhynocoris marginatus* for several generations without supplying any insect prey and showed a good nymphal survival rate. For the first time, initially Sahayaraj et al. (2006) developed oligidic diet for rearing this bug based on the comparison of natural and factitious preys.

Factors which influence the consumption and growth rates of insects were protein, sucrose, and water (Martin and Van't Hof 1988; Timmins et al. 1988; Karowe and Martin 1989; Slansky 1993). Shifting of nutritional requirements and capabilities is a general feature of heteropteran nymphal development as most reduviid bugs substantially change in body size, ecology, and physiology across nymphal instars (Avila et al. 2003).

Functional goals should change from early instars acquiring nutrients necessary for additional growth to the late instars for metamorphosis, reproduction, and adulthood. In this chapter, we considered practical aspects of the feeding behavior of reduviids against artificial diet. Further, the feeding behavior relies upon direct observation and note taking.

2.3.1 Consistency of Artificial Diet

In general liquid, semiliquid, gel, and solid form of artificial diets were prepared and utilized for hemipteran predator rearing. As a member of the Hemiptera, *Rhynocoris marginatus* is said to be a liquid feeder. Reduviid biologists over five decades used variety of natural hosts, laboratory hosts, to rear reduviid predators.

Further, reduviid predators are having piercing and sucking type of mouthparts (rostrum). The rostrum has long tubes with small opening in the front. Hence the fluid moves into the tube very easily into the esophagus and then to the crop. But, these insects, along with many other predatory hemipterans, target solid foods, which they liquefy with digestive enzymes (Sahayaraj and Muthukumar 2011; Kumar and Sahayaraj 2012) and specialized mechanical action. However, a solid diet is also unsuitable for rearing sucking type of mouthpart predators because it can easily become contaminated and dried out (Thompson and Hagen 1999; Tan et al. 2015). Hence we confined to formulate a liquid diet described in the presiding section. Various analytical approaches were employed to try to determine the different requirements and define the food composition for entomophages. For predators, the presentation of the diet is a key parameter although the respiration is not mainly involved.

Liquid diets were fully or semi-defined and were presented within wax capsules for the neuropteran *Crysoperla carnea* and the hemipteran predator *Geocoris punctipes*. Diets for different lacewing species (chrysopids) were encapsulated, presented on cellulose sponge or in the form of a free hygroscopic powder. Diets for predaceous coccinellids could be presented in gelled cubes or as powder, or dry pellets.

Stretched Parafilm was used to package diets with a paste-like consistency for several hemipteran predators, as well as for some species of coccinellids. Parafilm enclosing synthetic foam cubes soaked with diet devoid of insect components was successfully used to rear *Macrolophus caliginosus* (Grenier et al. 1989).

2.3.2 Meridic Artificial Diet Ingredients

The results are highlighted in this chapter. According to De Clercq (2004), the artificial diet has been categorized to three different types:

1. Holidic diets, where ingredients are known in chemical structures.
2. Meridic diets have a holidic base supplemented with one or more unrefined or chemically unknown substance (e.g.: liver extracts, yeast products, etc.).
3. Oligidic diets are mainly made up of crude organic materials (like meat diets).

Earlier work from our laboratory confirmed the optimal proportions of the basic ingredients for a meridic diet based on effects of the development and reproductive capacity of *Rhynocoris kumarii* (Sahayaraj et al. 2006). Although a promoted liquid diet could be easily ingested by sucking in *Rhynocoris kumarii*, it led to high predator mortality because the liquid feed was somewhat nutrient deficient. Moreover, the liquid forms of previously attempted artificial diet did not meet the requirements in terms of development and reproduction of reduviids based on the developed meridic diet using cotton ball for the mass rearing of *Rhynocoris marginatus*. Previous studies showed that meat-based substances, such as beef and pork liver, had a prominent role in the artificial diet in terms of the growth of reduviids (Sahayaraj et al. 2006).

Similarly, artificial diets used in the mass rearing of reduviids that included pork blood, meridic egg yolk, and protein-X were known to be metabolized for nutrition and energy to the reduviids (Sahayaraj et al. 2006). In addition, chemical substances, such as sodium chloride and potassium chloride, were included as necessary components of the diet. The aim of the present chapter is to determine the optimal meridic diet for *Rhynocoris marginatus* based on the developmental and reproductive status of the predator and also investigate the influence of diet on *Rhynocoris marginatus* compared with when fed with natural prey.

In 2002, we developed two meridic diets [diet 1 and diet 2] based on *Spodoptera litura* [diet 1] and *Bombyx mori* [diet 2] whole body-based diet and recorded the feeding behavior of *Rhynocoris marginatus*. The ingredients are listed below:

Ingredients	Diet 1	Diet 2
Insect whole body	*Spodoptera litura*	*Bombyx mori*
Beef liver (g)	3.0	3.0
Milk powder (g)	2.0	2.0
Honey (ml)	15	15
Acetic acid –10 % (ml)	2.7	3.7
Egg yolk (g)	2	2
Sucrose (mg)	200	200
Vitamin complex (mg)	40	40
Streptomycin (mg)	25	25
NaCl (mg)	5	5
KCl (mg)	5	5
Protein-X (mg)	250	250
Water (ml)	100	100

Along with the insect ingredients, we also tested beef liver-based diets against this reduviid. Various ingredients are as follows.

Ingredients	Diet 3	Diet 4
Beef liver (g)	10.0	–
Pork liver	–	10.0
Milk powder (g)	2.0	2.0
Egg yolk (mg)	20	20
Sucrose (mg)	200	200
Ascorbic acid (mg)	10	10
Wesson's salt (mg)	20	20
Water (ml)	100	100

The diet is devoid of insect ingredients, milk powder, protein-X, and individual salts but rather added with Wesson's salt mixture along with ascorbic acid. The diets 3 and 4 consist of the following ingredients in different proportions.

Ingredients	Diet 5	Diet 6
Beef liver (g)	30.0	60.0
Honey (ml)	2.5	5.0
Acetic acid –10 % (ml)	1.85	3.7
Egg yolk (g)	2.0	4.0
Sucrose (mg)	100	200
Vitamin complex (mg)	5.0	10.0
Streptomycin (mg)	3.75	7.5
NaCl (mg)	2.15	5.0
KCl (mg)	2.15	5.0
Water (ml)	100	100

In diets 5 and 6, milk powder and protein-X were completely omitted as in diets 3 and 4. Similarly, the other component quantities were reduced, due to the addition of sole components of the beef liver. These diets support the development of nymphal instars of *Rhynocoris marginatus*, but failed to support the reproduction. Biochemical analysis of the diet reveals that diet 5 consists of 0.46, 0.34, and 0.23 mg/ml total carbohydrate, protein, and lipid, respectively. Diet 6 contains more total carbohydrate (0.6 mg/ml), total protein (0.46 mg/ml), and total lipid (0.29 mg/ml).

It was recorded that diet contaminations were common when the diet was devoid of formaldehyde. After the addition of formaldehyde, nearly 70 % of diet contaminations have been prevented. Indirectly, it also prevented the *Aspergillus flavus* infection to *Rhynocoris marginatus* nymphs and adults. Furthermore, the diet 5 and diet 6 devoid of yeast may also prevent the contamination and infection. Our results also showed that yeast extract provides a substantial quantity of these factors, since addition of yeast extract to the base diet promotes predator's development and also the reproduction (diet 5). The yeast extract was more effective for increasing the number of adult emergence of *Rhynocoris marginatus* than the liver diet (Xie et al. 1997). Meat diet produced smaller, lighter-weight insect with longer embryonic and nymphal developmental period than the control categories. This common feature is discussed by Castane and Zapata (2005) and Grenier and De Clercq (2003).

Experience of 2003, during 2006, it developed a series of diets [diet 7 to diet 12] with beef liver along with pork blood and pork blood serum. This meat-based meridic diet is prepared following the method of Sahayaraj et al. (2006) with some modifications. The source and other ingredients present in the oligidic diets are presented in Table 2.1.

Artificial diet contains a nitrogen source (protein/free amino acids), lipids, carbohydrate, water-soluble vitamins (vitamins B and C) and lipid-soluble vitamins (vitamin A), minerals, preservatives, and antimicrobial protective agents.

Both Na and K are involved in regulation of pH in the cells and body fluids (hemolymph, salivary secretion) of insects. Since minerals cannot be synthesized by insects, it included in the diet. Chloride is universally required for organisms. Insects are involved in the membrane potential and enzyme reactions. Further,

Table 2.1 Composition of various ingredients of oligidic diets (for 100 ml)

	Diets					
Components (in mg/ml)	Diet 7	Diet 8	Diet 9	Diet 10	Diet 11	Diet 12
Source ingredient (g)	60	5	5	5	5	5
Pork blood (g)	–	–	–	5	5	–
Blood serum (ml)	–	–	–	–	20	20
Sucrose (mg)	200	200	–	500	500	500
Yeast extract (water soluble)	–	–	–	–	5	5
Yeast (g)	–	–	2	2	–	–
Milk powder (g)	–	–	5	5	5	5
Egg yolk (g)		4.0	4.0	4.0	5	5
Honey (ml)	5.0	5	5	5.0	5.0	5.0
Vitamin (multivitamin mg)	10.	200	200	200	200	300
Vitamin E	–	–	–	200	200	200
Vitamin C	–	–	–	2.5	2.5	2.5
Casein (mg)	–	–	5	4	4	4
Cholesterol (mg)	–	–	200	200	200	200
Acetic acid (10 %) (ml)	3.7	–	–	2.5	2.5	2.5
NaCl (mg)	5.0	5.0	5.0	–	–	–
Streptomycin (mg)	7.5	100	100	100	–	–
Formaldehyde 40 % (ml)	–	–	–	–	1	1

minerals along with other nutrients (sugar, some amino acids, lipids, and ascorbic acids) act as feeding stimulators. Mineral salts added also do contribute to the nutritional value.

Since streptomycin has been used as antibiotic in humans, it is used the same as antibacterial agent in the diet as suggested by Hsiao (1979). It gives a great amount of promise toward antimicrobial therapy for insect too.

2.3.3 Preparation

Source ingredients such as pig liver and pig blood were dried in hot air oven at 60 °C for 25–30 min. They were ground well by mortar and pestle and stored in refrigerator for the use of month. Hundred milliliters of distilled water was boiled at 100 °C for 20 min. Milk powder (Lactogen, Nestle, Mumbai, India) was dissolved in 10 ml of boiled water and it was allowed to cool. Water-soluble yeast extract (Fine Chemicals Ltd., Mumbai), dried egg yolk, liquid honey (Dabur, Narendrapur, West Bengal, India), and acetic acid (Glaxo, Gujarat, India) were added to the remaining 90 ml water and in desired quantity. After 10 min, the temperature was reduced to 40 °C, and then the source ingredients, multivitamin, vitamins C and E, and streptomycin (Sarabairaman, Vadodara, India) were added and stirred for thorough mixing. Then the milk powder solution was added and stirred well. After the thorough mixing, the prepared diet was allowed to cool at room temperature and

Plate 2.1 Liquid diets predated by using insects and vertebrate diver as source ingredients

then it was filtered through Whatman no. 1 filter paper to obtain a liquid diet (Plate 2.1). Filtered liquid diet was stored in 125 ml reagent bottles in refrigerator for the future use on longer than 2–3 weeks.

For the practical conveniences, it stored the diet in refrigerator assuming that storage will not significantly degrade the diet materials. It was also realized that during the course of our study, none of the ingredients and diets change after the storage. However, the addition of excess yeast showed characterized smell in the diet which had been stored for a couple of weeks together.

It has prepared more than 17 diets. But it listed out the composition/constituents of the meridic diets which are in favor for the development and fecundity of *Rhynocoris marginatus*. All the listed diets contain pork liver as a source ingredients. Other ingredients were present in different concentrations at different proportions.

2.3.4 Object Preference

This study is particularly useful in determining the object choice, helping to determine which object that the predator is capable of locating feeding the artificial diet. Initially we tested four objects, namely, cotton, paraffin capsule, commercial foam, and cavity microslide. Proportion of predator approaching and feeding the diet with specific object, approaching time, and consumption time were considered for preference. Initially we tested four objects (cotton, capsule, cavity slide, and foam) for providing the OD to *Rhynocoris marginatus* of different life stages.

First stadium (48.5 %), fifth stadium (33.3 %), and adult predators (50.5 %) preferred to feed the artificial diet provided in cotton. Furthermore, adults took 3.3 and 7.1 min to approach and consume the food present in cotton, respectively, whereas fifth nymphal instars and first nymphal instar took 5.1 and 7.2 min, respectively, with a consumption time of 4.9 and 3.13 min. All life stages easily feed the artificial diet present in the cotton either alone or in group (Plate 2.2). They preferred cotton (Plate 2.2b) and hence reduviid easily inserted in rostrum and suck the artificial diet without any constrains. Further coastwise it was lowest price than other three objects tested.

Parafilm capsule is another object preferred by the reduviids (33.3 %, 23.3 %, and 25.5 % for first stadium, fifth stadium, and adults, respectively) (Plate 2.2). Feeding time varied with life stages of *Rhynocoris marginatus* (2.89, 1.73, and 2.44 min for first stadium, fifth stadium, and adults, respectively). The observation reveals that the Parafilm disintegrated after some time of feeding and reduviid feeds alone or in groups as it follows when it encountered an insect prey. Further, Parafilm color does not have any influence on capturing time.

2.4 Feeding Behavior on Meridic Diet

The following feeding behaviors were recorded while meridic diet was provided to the reduviids:

1. *Reduviid moving forward across the feeding arena.*
2. *Motionless standing.*
3. *Making rapid movements with fore- and hind legs across body surface and antennae.*

a

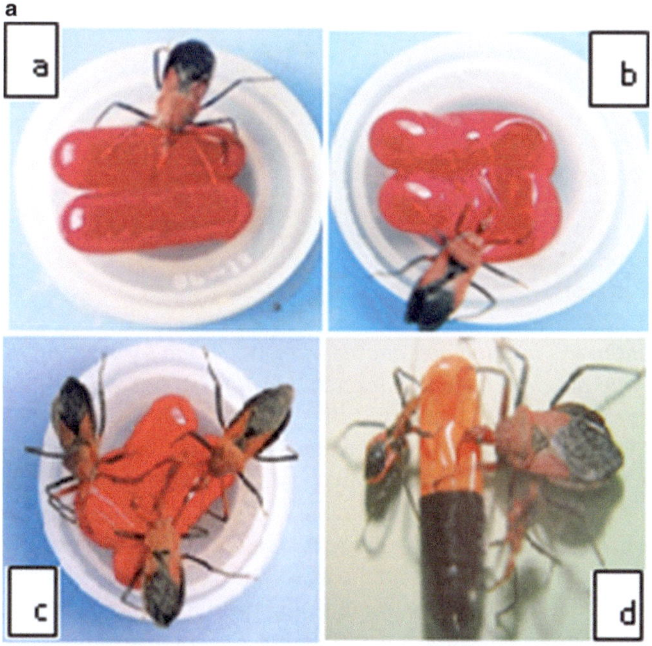

Plate 2.2 (**a**) Feeding acts of reduviid predator, *Rhynocoris marginatus* adults individually feeding on artificial diet (**a** , initial; **b**, at the end) and also in group in *red*-colored Parafilm (**c**) and *red*- and black-colored Parafilm (**d**). (**b**) Acts of *Rhynocoris marginatus* nymph (**a**) and adults [**b**, individually; **c**, in group], *Rhynocoris fuscipes* (**d**, approaching; **e**, feeding), *Rhynocoris longifrons* (**f**), and *Panthous bimaculatus* adult (**g**) feeding on artificial diet provided in cotton

Plate 2.2 (continued)

4. *Insert mouthparts into the diet packet, but not feeding for any length of time.*
5. *Insert mouthparts into the diet packet and feed for an extended amount of time.*

The abovementioned acts are common among the reduviid predators. This indicates chemicals as well as other parameters like texture, consistency, etc. It attracted the reduviids to feed the meridic diet. Preliminary behavioral studies were conducted in reduviids to understand their approaching behavior toward the meridic diets. Field-collected and laboratory emerged adults as well as nymphs of *Rhynocoris marginatus* were used for the study. In general approaching time decreased when the predator grew older. For instance, first stadium *Rhynocoris marginatus* took only 1.4 min to feed on *Bombyx mori*-based artificial diet and took 5.4 min when it reached the fifth stadium. During feeding *Rhynocoris marginatus* try to insert in rostrum deepens into the cotton (the feeding arena). However, it was not true for all diets tested here. Another observation we recorded is nymphs took more time to consume artificial diet, whereas adults took less time to consume the same which reflects that adults' metabolic activity is higher than nymphs.

Pork liver-based diet (diet 3) was fortified with protein-X to enhance the protein level of the artificial diet. The fortification was done in four concentrations as 5, 10, 15, and 20 %. Based on the food consumption time (minutes), we concluded that incorporation of 5 % protein X (w/v) enhanced the feeding time of *Rhynocoris marginatus* nymphs (16.59 and 22.24 min for first and second stadium, respectively) than adults (16.24 min).

Water is an important component of the meridic diet; hence it tested water against the artificial diet. Results reveal that *Rhynocoris marginatus* consumed more amount of artificial diet than water. It supports the idea of Edney (1977) and Chapman (1998) saying that many insects feed on water to sustain their life. Correlation was made between water consumption and artificial diet 1, 2, 3, and 4 consumption, and very positive correlation coefficients were recorded ($r^2 = 0.85$, 0.72, 0.68, and 0.41 for diets 2, 3, 1, and 4, respectively).

2.4.1 Feeding Behavioral on Meridic Diet in Relation to Starvation

Starvation influences the approaching time, sucking time, and weight gain of *Rhynocoris marginatus* male and female when provided with artificial diet. Animal weight was gradually diminished while the predator grew older (Table 2.2). Males approached the diet more quickly than did by females. At 24 and 72 h starvation female consumed more food, because they spent more time to consume the food. But this was not true when this predator was starved for 48 h (Tables 2.2 and 2.3).

Instead of 10 g of beef liver, in diets 5 and 6, 30 and 60 g of beef liver were incorporated, respectively, and the feeding activity of *Rhynocoris marginatus* nymphs and adults was recorded (Sahayaraj and Sujatha 2003 unpublished data) in relation to 24, 48, 72, and 96 h starvations. In second, third, fourth, and fifth stadium, the consumption time gradually increased, while the food deprivation period prolonged from 24 to 96 h both at 30 and 60 g beef liver added in diets (Table 2.5). Calculated

Table 2.2 Feeding behavior of field-collected adult *Rhynocoris marginatus* on artificial diets

Sex	Animal weight (mg)	Approaching time (M)	Sucking time (minutes)	Weight gain (mg)
24 h starvation				
Female	199.2 ± 11.7	1.2 ± 0.9^{NS}	59.6 ± 9.7*	30.6 ± 4.8*
Male	110.3 ± 12.4	5.6 ± 2.1	06.5 ± 3.0	08.1 ± 2.5
48 h starvation				
Female	197.1 ± 8.3	12.7 ± 3.0 ^{NS}	11.8 ± 2.4*	24.2 ± 2.8*
Male	102.8 ± 7.1	08.5 ± 2.1	12.3 ± 2.3	06.3 ± 3.2
78 h starvation				
Female	151.4 ± 0.5	5.5 ± 1.7*	26.1 ± 0.3*	21.0 ± 2.8*
Male	085.0 ± 0.5	2.7 ± 0.6	20.8 ± 0.1	04.7 ± 1.1

* indicates significance at 5 % level. NS indicates not significant

Table 2.3 Feeding time (minutes) of *Rhynocoris marginatus* of different life stages starved for 24, 48, 72, and 96 h and then provided with artificial diet prepared using 30 and 60 g beef liver as source ingredients

		Starvation time (hours)			
Life stages	Beef liver quantity (g)	24	48	72	96
First stadium	30	17.7	2.8	7.6	12.1
	60	6.1	6.2	4.4	6.3
Second stadium	30	2.6	4.4	8.8	10.5
	60	2.6	4.0	6.3	14.1
Third stadium	30	1.5	3.7	7.0	10.1
	60	1.1	4.7	5.0	15.6
Fourth stadium	30	2.3	3.3	5.3	14.5
	60	0.6	2.6	3.6	6.8
Fifth stadium	30	2.3	6.3	7.7	7.9
	60	2.8	4.8	5.0	5.8
Adult	30	–	–	6.1	–
	60	–	–	5.2	–

– indicates no observation was recorded

Access Proportion Index (API) showed that it was higher (API = 0.8–1.0) both in first and second stadium than in third, fourth, and fifth stadium (API = 0.6–1.0).

As reported by Hill (1989), sugar is a very important component which promotes the egg production. Similarly McEwen and Kidd (1995) had recommended yeast and sugar for maximum egg production. Honey is also a very important component regarding fecundity. McEven and Kidd (1995) and Kubota and Shiga (1995) analyzed that a mixture of honey and yeast autolysate is a suitable adult diet for production of fertile eggs. Reproduction of *Rhynocoris marginatus* from the diets 4–6 and the non-reproduction in other diets must not be due to the honey. Sahayaraj et al. (2006) reared *Rhynocoris marginatus* adult with artificial diet consisting of milk, eggs, sugars, and yeast which is found to be in favor for fecundity. Higher fecundity was observed in diet containing higher egg yolk (amino acids 15.5 %) as observed by Norioka et al. (1984). Moreover, *Rhynocoris marginatus* reproduced after the addition of vitamins E and C, blood serum, casein, cholesterol, and acetic acid. Hence it hypothesized that all these constituents are essential for this reduviid reproduction and development.

2.4.2 Choice and Non-choice Test Against Artificial Diet

The suitability of the oligidic diets and feeding behavior of *Rhynocoris marginatus* were determined by both the choice and non-choice tests. For the laboratory testing and 1–3 concentrations of the artificial diets (1–3) were placed in the olfactometer along with water. Then 24 h starved *Rhynocoris marginatus* adults were introduced into the glass olfactometer. Fifteen replications were made for each experiment. The approaching time (AT) and consumption time (CT) and choice of feeding by

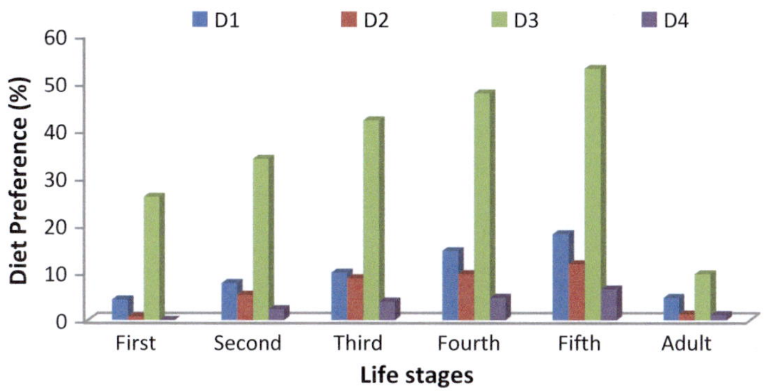

Fig. 2.1 Preference of *Rhynocoris marginatus* different life stages against *Spodoptera litura*-based diet (diet 1), *Bombyx mori*-based diet (diet 2). Beef liver-based diet (diet 3) and pork liver-based diet (diet 4)

Fig. 2.2 Pie diagram showing the selection of artificial diet fed to *Rhynocoris marginatus* of different life stages collected from laboratory stock

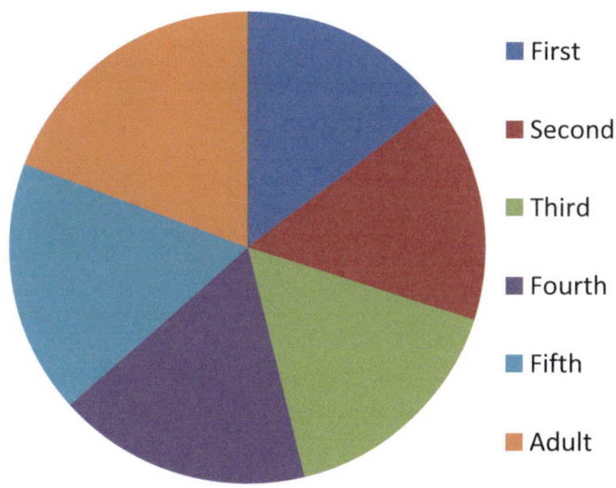

reduviid were recorded for 1 h continuously by visual method in each experiment. Invariable life stages preferred beef liver-based diet (diet 3) followed by *Spodoptera litura*-based diet (diet 1). This was assessed based on percent of time spent by the predator to feed the diet. Though insect diet is present in diets 1 and 2, the predator preferred the diet which contains beef liver as source ingredient (Fig. 2.1). It is a good sign showing that mass production of reduviid predator can be done by utilizing vertebrate body parts as a source ingredient. Our further research was carried out by us, based on these findings.

2.5 Best Strain Selection

Both nymphs and adults of *Rhynocoris marginatus* collected from the laboratory stock culture were provided with laboratory hosts and artificial diet. More than 72 % of adults preferred artificial diets followed by fifth stadium, fourth stadium, third stadium, second stadium, and first stadium (Fig. 2.2).

Table 2.4 Stage preference of *Rhynocoris marginatus* on *Corcyra cephalonica* (CC), *Spodoptera litura*, and *Dysdercus cingulatus*

	Preys		
Predator life stages	*D. cingulatus*	*S. litura*	*C. cephalonica*
II	II	II	II
III	II	II	II
IV	Adult	III	IV
V	Adult	IV	V
Male	Adult	V	V
Female	Adult	V	V

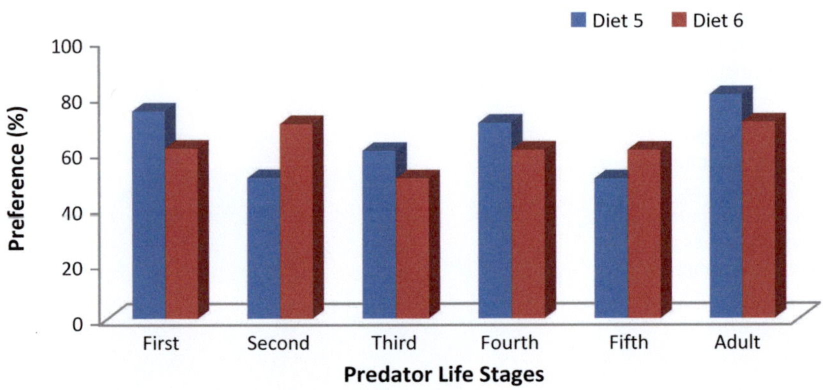

Fig. 2.3 Bar diagram showing the selection of artificial diets 5 (30 g beef liver) and 6 (30 g beef liver) fed to *Rhynocoris marginatus* of different life stages collected from laboratory stock

In another study *Rhynocoris marginatus* of different life stages (nymphs and adults) were provided with diets 5 (30 g beef liver) and 6 (30 g beef liver) as against *Corcyra cephalonica*. Except second stadium, all other life stages highly selected diet 5 (Table 2.4; (Fig. 2.3)) (Sahayaraj and Sujatha 2003 unpublished data).

Influence of diet on the predation of *Rhynocoris marginatus* on natural preys.

2.6 Insect Prey Preference After Fed with Meridic Diet

2.6.1 Stage Preference of Reduviid Fed with Artificial Diet

Stage preference study of *Rhynocoris marginatus* of different life stages fed with *Dysdercus cingulatus*, *Spodoptera litura*, and *Corcyra cephalonica* separately was carried out with experiment of choice as described by Holling (1966). *Rhynocoris marginatus* was introduced into a Petri dish and *Dysdercus cingulatus* first, second, third, fourth, and fifth instar nymphs were released, and the predatory behavior was observed consecutively for 6 h visually. The stage where preys were successfully

Plate 2.3 *Rhynocoris marginatus* female feeding on adult *Dysdercus cingulatus*

captured, killed, and consumed was recorded as the preferred stage of the reduviid. Fourth, fifth, and adult predators were provided for the third, fourth, fifth nymphal instars and adults of *Dysdercus cingulatus*. For both *Spodoptera litura* and *Corcyra cephalonica* preference, life stages of *Rhynocoris marginatus* were provided with all the five nymphal instars of the prey separately. Ten replications were maintained for each life stage of the predator separately. The preferred life stages of the pests were used for the biological control potential evaluation studies.

Reduviid encountered different developmental stages of the host within a patch. Those stages potentially vulnerable to attack may differ in their profitability. Stage preference studies of *Rhynocoris marginatus* of different life stages fed with *Spodoptera litura*, *Corcyra cephalonica*, and *Dysdercus cingulatus* are presented in Table 2.4. Prey stage preference studies showed that life stages of *Rhynocoris marginatus* preferred to different stages of the pests tested. The result also suggests that both fifth instar and adult predators were successful in encountering the large size *S. litura* (iv and v instars) and *Corcyra cephalonica* (v instars) larvae. Though the different nymphal instars of *Rhynocoris marginatus* preferred Lepidoptera larvae, second and third instar reduviids preferred second instar *Dysdercus cingulatus* nymphs, and the reduviid belonging to the remaining life stages preferred adult *Dysdercus cingulatus* (Plate 2.3).

2.6.2 Food Preference Index (FPI)

The artificial diet was provided with additional combination with different foods such as oligidic diet (T3), oligidic diet + *Corcyra cephalonica* (T4), and *Corcyra cephalonica* + water (T2). One ml of OD was provided once in 2 days. After 2 days, cotton ball was removed and discarded in order to maintain the hygienic conditions. Moreover, before the insects were introduced in the container (100 and 500 ml capacity), they were washed with 0.2 % sodium hypochlorite. Thirty male and female *Rhynocoris marginatus* from each treatment were randomly selected, weighed, and introduced to olfactometer covered with muslin cloth. Among the tested insects, female took more time (59.6 min) for sucking and also gained more (30.6 mg) weight in all the tests (Table 2.5).

Table 2.5 Cumulative approaching behavior of field-collected adult *Rhynocoris marginatus* on oligidic diets

Sex	Animal weight (mg)	Approaching time (M)	Sucking time (minutes)	Weight gain (mg)
24 h starvation				
Female	199.2 ± 11.7	1.2 ± 0.9^{NS}	59.6 ± 9.7*	30.6 ± 4.8*
Male	110.3 ± 12.4	5.6 ± 2.1	06.5 ± 3.0	08.1 ± 2.5
48 h starvation				
Female	197.1 ± 8.3	12.7 ± 3.0 ^{NS}	11.8 ± 2.4*	24.2 ± 2.8*
Male	102.8 ± 7.1	08.5 ± 2.1	12.3 ± 2.3	06.3 ± 3.2
78 h starvation				
Female	151.4 ± 0.5	5.5 ± 1.7*	26.1 ± 0.3*	21.0 ± 2.8*
Male	085.0 ± 0.5	2.7 ± 0.6	20.8 ± 0.1	04.7 ± 1.1

* indicates significance at 5 % level. NS indicates not significant

2.6.3 Feeding Behavior with Live Preys

The laboratory emerged *Rhynocoris marginatus* life stages (except first instar) were used for evaluation of their biocontrol potential. Preferred stages of *Dysdercus cingulatus* (5/container) were introduced into the container containing cotton twig, and it was allowed to acclimatize for 10 min. Then a *Rhynocoris marginatus* was introduced in the same container, and the feeding events like approaching time, handling time, and sites preferred by the reduviid for feeding were recorded continuously for 6 h. After 24 h, weight is gained and number of prey consumed by a predator. It was recorded. Ten replications were maintained for each life stage of the predator separately.

2.6.3.1 Approaching Time (AT)

Field-collected reduviids were used to study the feeding behavior on oligidic diets. At 24 h starvation, field-collected *Rhynocoris marginatus* adults oriented toward the oligidic diets. At 48 h of starvation, field-collected adults also exhibited 30 % responses toward milk-based oligidic diets, and then it was increased to 60 % approaching within 1 h. At 72-h hunger level, 70 % of the predators approached the diet. When 5 % of insect sources such as *Mylabris indica*, *Mylabris pustulata*, and *Dysdercus cingulatus* were added in the milk-based diet, reduviids failed to approach the diet within 2 h. During feeding, sequential events were observed: antennal stretching, cat walking toward the direction of diet, restless movement resulting in flight antenna brushing and leg brushing, and repeated movement toward the diet source. But the predators failed to approach the cotton containing milk-based diet with insect source. Further, from Table 2.3, it was very clear that *Rhynocoris marginatus* food consumption was gradually decreased when the starvation period was increased from 24 to 72 h. Similarly approaching and sucking time were also decreased. Between the two sexes, female consumed more milk-based oligidic diet than male. Significantly female spent more time to suck the diet than male.

The following is the sequential pattern of predatory behaviors (see Sahayaraj et al. 2007 for details):

1. *Arousal – the predator closely watches the movement of the prey, is restless, and has straightened legs.*
2. *Approach – the predator moves toward the prey, pointing the rostrum and antennae forward.*
3. *Capture – the predator adapts rate of movement to that of the prey; the prey is closed and captured held tightly by predator's forelegs.*
4. *Rostral probing – the predator probes at various parts of the body of the prey with the rostrum.*
5. *Paralyzing the prey – the predator draws back rostrum; the prey loses agility and becomes immobile indicating that paralysis has occurred.*
6. *Feeding – the predator transports the prey to a secluded place and sucks out the body content and inserts its rostrum at one or more places.*
7. *Post-predatory behavior – the predator drops the carcass and cleans the rostrum, antennae, and forelegs.*

The predator time used on each of the first three events (1–3) was summed up and used as a collective measure of locating and capturing the prey (termed approaching time, AT). Similarly, the time that the predators used on each of the subsequent three events (4–6) was summed up and used as a collective measure of handling and eating the prey (termed handling time, HT). Our observations reveal that during the feeding time, *Rhynocoris marginatus* oriented toward the prey with facing antenna, after setting a perfect orientation position, the reduviid palpated antenna then aroused subsequently showed the approach rostral probing, injection of toxic saliva for paralyzing, sucking the prey content, and post-predatory behavior observed in this study.

The approaching behavior was also recorded for *Rhynocoris kumarii*, *Panthous bimaculatus*, and *Sycanus collaris* different life stages fed with meat-based artificial diet. Results show that reduviid approaches the diet with the antennae forward (Sahayaraj 2012), which suggests a visual and chemical recognition of the diet as observed for *Rhynocoris marginatus* (Fab.) (Sahayaraj et al. 2006, 2007; Sujatha and Sahayaraj 2007) during feeding.

2.6.3.2 Capturing

At close proximity of the prey, the predator extended rostrum and captured the prey preferably in the abdominal region. If the prey is small and less active, the predator captured the prey with forelegs firmly kept over the prey. If the prey was agile, the predator used to raise antennae and extended rostrum and pinned the prey at a preferred site (Table 2.6). The predator was found probing the motionless prey with antennae presumable test the prey with inserted rostrum. But the acts did not affect the subsequent predatory activities such as pinning and paralyzing. The extended rostrum was inserted into the captured prey to test the suitable site for sucking. The fed predators took more time for capturing and pinning when compared with prey-deprived predators.

Table 2.6 Site preference of *Rhynocoris marginatus* during feeding on three pests

Predator life stages	*Dysdercus cingulatus*	*Spodoptera litura*	*Corcyra cephalonica*
II	Thoracic pleural membrane	Thoracic pleural membrane, tergum	Sternum, neck membrane
III	Eye, thoracic pleural membrane	Thoracic pleural membrane, tergum	Neck membrane, sternum
IV	Thoracic and abdominal pleural membrane	Thoracic and abdominal pleural membrane	Neck and abdominal pleural membrane
V	Thoracic and abdominal pleural membrane	Thoracic pleural membrane, tergum	Neck and abdominal pleural membrane
Male	Head neck muscle, tergum	Thoracic and abdominal pleural membrane	Neck and abdominal pleural membrane
Female	Neck membrane, tergum	Thoracic pleural membrane, tergum	Neck and abdominal pleural membrane

2.6.3.3 Paralyzing

After the successful capturing of the prey, the predator paralyzed the prey of injecting its toxic salivary secretion. The fed predators took more time for paralyzing the prey when compared to the prey-deprived predators.

2.6.3.4 Rostral Probing and Sucking

After the paralyzing of prey, the predator is sucked the predigested body fluids of the prey by inserting the rostrum at different regions of the prey body. *Rhynocoris marginatus* also frequently changed the sucking sites during feeding. While sucking one of the antennae was kept toward the prey in an upright position and the other antenna toward the prey in a droop and extended position, presumably to test various rations of the prey. It was further accomplished by rostral probing. The predators often selected the sucking sites from the abdominal region and less often from the cephalic and thoracic regions (Table 2.6). At the time of sucking, the forelegs of the predator were found kept on the prey. The fed predator took less sucking time, when compared to the prey-deprived predators. Lower prey ingestion might be due to the sifting of the consumption of available artificial diet.

2.6.3.5 Post-feeding Behavior

The satiated predator after sucking the prey at all possible sites started cleaning the rostrum and antennae, by drawing and grooming in between the forelegs followed by cleaning the fore tibial pad, antenna, and hind tibia.

2.6.4 Feeding Behavior Against Three Pests

First attacking site of *Rhynocoris marginatus* on three pests is presented in Table 2.6. It was very clear from Tables 2.7, 2.8, and 2.9 that reduviid of different life stages invariably preferred a particular site for paralyzing and sucking the prey content. Both paralyzing and feeding the victim reduviids are significantly are prefer

Table 2.7 Feeding behavior (in min) and weight gain (in mg) of *Corcyra cephalonica* (CC)- and artificial diet-reared *Rhynocoris marginatus* on *Dysdercus cingulatus* of different life stages

Instar	Approaching time	Handling time	Weight gain	No. of sites selected
Dysdercus cingulatus				
II	7.0 ± 1.7	131.3 ± 9.5*	08.1 ± 0.7	2.0 ± 0.2
III	4.7 ± 0.7	167.2 ± 9.9	18.1 ± 1.1*	2.3 ± 0.1
IV	4.4 ± 0.6	199.5 ± 10.3	31.0 ± 4.5*	2.4 ± 0.2
V	1.7 ± 0.2	141.6 ± 8.3	41.5 ± 0.0	2.6 ± 0.2
Male	4.1 ± 0.9	128.5 ± 3.7	25.6 ± 3.1	1.8 ± 0.3
Female	5.0 ± 1.4	138.7 ± 2.5	35.2 ± 2.8*	1.6 ± 0.2
Artificial diet				
II	7.3 ± 0.8	88.0 ± 08.6	1.9 ± 0.0*	11.6 ± 0.1*
III	7.3 ± 1.7	259.5 ± 8.0*	11.1 ± 1.6	3.2 ± 0.2*
IV	3.5 ± 0.5	282.8 ± 24.3*	15.8 ± 3.2	2.2 ± 0.2
V	4.3 ± 1.0	214.5 ± 12.2*	31.2 ± 4.6*	2.3 ± 0.1
Male	4.3 ± 0.8	144.2 ± 26.4*	35.2 ± 4.3*	3.2 ± 0.5*
Female	5.4 ± 0.7	125.0 ± 20.8	25.6 ± 3.1	2.6 ± 0.5*

*Significant at 5 % level

Table 2.8 Approaching time, handling time (in min), and weight gain (in mg) of *Rhynocoris marginatus* of different life stages on *Corcyra cephalonica* and oligidic diet

Predator stage	Approaching time	Handling time	Weight gain	No. of sites selected
C. cephalonica				
II	2.2 ± 0.4	130.6 ± 29.1*	7.3 ± 9.0*	1.3 ± 0.1*
III	2.0 ± 0.4	120.0 ± 25.5*	11.7 ± 0.3*	1.1 ± 0.0
IV	1.5 ± 0.2	134.8 ± 11.3*	11.4 ± 0.2*	2.1 ± 0.2*
V	1.4 ± 0.2	157.2 ± 09.1*	22.9 ± 8.2*	2.2 ± 0.2*
Male	1.3 ± 0.2	165.0 ± 20.4*	34.3 ± 3.5	1.8 ± 0.3*
Female	1.6 ± 0.4	163.3 ± 20.6*	33.2 ± 0.3*	1.6 ± 0.2*
Artificial diet				
II	4.6 ± 1.0	227.5 ± 14.4*	8.0 ± 0.6*	1.5 ± 0.1
III	4.7 ± 0.8	215.0 ± 18.1*	20.5 ± 2.4*	1.8 ± 0.1*
IV	4.4 ± 0.8	250.1 ± 10.3*	48.1 ± 2.8*	1.6 ± 0.1
V	3.0 ± 0.8	270.5 ± 12.9*	43.9 ± 8.6*	1.2 ± 0.1
Male	4.6 ± 0.8	159.0 ± 19.9*	34.6 ± 6.5	2.2 ± 0.2*
Female	4.0 ± 1.6	198.0 ± 21.9*	53.0 ± 9.8*	2.4 ± 0.2*

*Significant at 5 % level

the thoracic pleural membrane of the pests followed for the abdominal pleural membrane. In general approaching time gradually diminished from the second instar to fifth nymphal instars (7.0, 2.2, and 2.4 min for *Corcyra cephalonica*, *Spodoptera litura*, and *Dysdercus cingulatus*, respectively) (Tables 2.7, 2.8, and 2.9). Meridic

Table 2.9 Approaching time, handling time (in min), and weight gain (in mg) of *Rhynocoris marginatus* of different life stages on *Spodoptera litura* and artificial diet

Life stage	Approaching time	Handling time	Weight gain l(in mg)	No. of sites selected
Spodoptera litura				
II	24.3 ± 0.8	126.5 ± 9.9*	8.6 ± 0.7*	1.6 ± 0.1*
III	19.5 ± 2.4	148.0 ± 8.7*	19.0 ± 2.2*	1.8 ± 0.1
IV	13.7 ± 2.7	129.0 ± 9.8	21.5 ± 2.3	2.0 ± 0.2*
V	10.1 ± 1.6	159.0 ± 21.0*	54.0 ± 7.6*	2.1 ± 0.2*
Male	11.2 ± 4.7	188.7 ± 18.7	34.5 ± 5.5	1.5 ± 0.2
Female	9.6 ± 2.0	175.0 ± 11.6*	29.0 ± 3.4*	1.5 ± 0.2*
Artificial diet				
II	7.4 ± 1.1	230.5 ± 11.1*	6.7 ± 0.5	2.1 ± 0.2*
III	4.5 ± 1.2	207.0 ± 11.2*	13.3 ± 1.2	1.8 ± 0.2
IV	3.3 ± 0.8	241.5 ± 13.1*	60.7 ± 1.9*	1.5 ± 0.6
V	3.8 ± 0.8	241.5 ± 13.1*	63.7 ± 1.9**	1.5 ± 0.1
Male	5.5 ± 1.6	225.0 ± 19.3*	63.7 ± 8.3*	2.0 ± 0.4*
Female	4.5 ± 1.2	227.5 ± 19.5*	76.1 ± 1.1*	1.8 ± 0.3*

*Significant at 5 % level

diet-reared *Rhynocoris marginatus* took more time for approaching the prey *Corcyra cephalonica* and *Dysdercus cingulatus* than natural host *Spodoptera litura*-reared predator. Statistical analysis between artificial diet-reared and prey reared-predator approaching times was insignificant at 5 % level. Artificial diet-reared reduviid approach is *Spodoptera litura* faster than other preys.

Irrespective of the preys, the oligidic diet-reared *Rhynocoris marginatus* significantly handled maximum time (270.5 min) than insect host-reared predator (Table 2.8). The results indicate that there was positive correlation between the handling time (250.10 min) and weight gain (48.10 mg). *Rhynocoris marginatus* second, third, fourth, and fifth instar nymphs and adult reared with insect hosts handled maximum time in *Dysdercus cingulatus* and *Spodoptera litura.* Minimum consumption time was observed in third instar *Rhynocoris marginatus* feeding on *Corcyra cephalonica* (Table 2.8). *Rhynocoris marginatus* female took more time for handling *Corcyra cephalonica* and *Spodoptera litura* than male. But male took less time when *Rhynocoris marginatus* was provided with *Dysdercus cingulatus* (Table 2.9). Irrespective of the preys, the weight gain was gradually increased when *Rhynocoris marginatus* grew older (except adult).

It is revealed from Tables 2.7, 2.8, and 2.9 that maximum weight gain was recorded in adult female when artificial diet-reared *Rhynocoris marginatus* was provided with *Spodoptera litura* (76.1 mg) (Table 2.8). But an opposite trend was observed for other pests, for instance, when *Dysdercus cingulatus* was provided to this reduviid predator. In general, artificial diet-reared *Rhynocoris marginatus* consumed more number of preys than those reared on *Corcyra cephalonica* and *Spodoptera litura.* Similarly, in artificial diet-reared *Rhynocoris marginatus*, maximum predatory rate was observed for *Spodoptera litura* adult female (2.40 prey/predator/day), and it was very low in third instar *Dysdercus cingulatus* (1.36 prey/

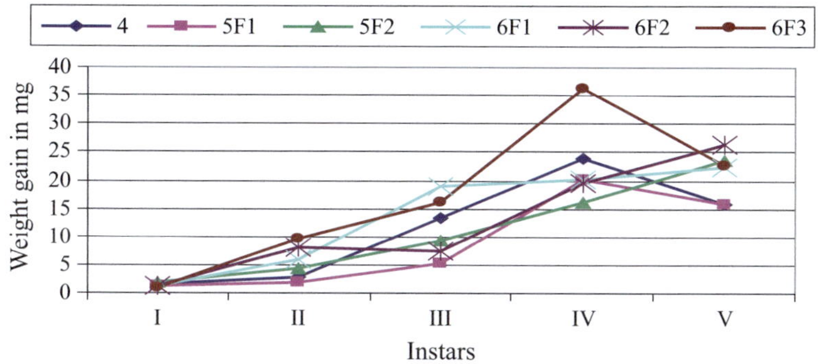

Fig. 2.4 Weight gain (in mg) of *Rhynocoris marginatus* reared on artificial diet

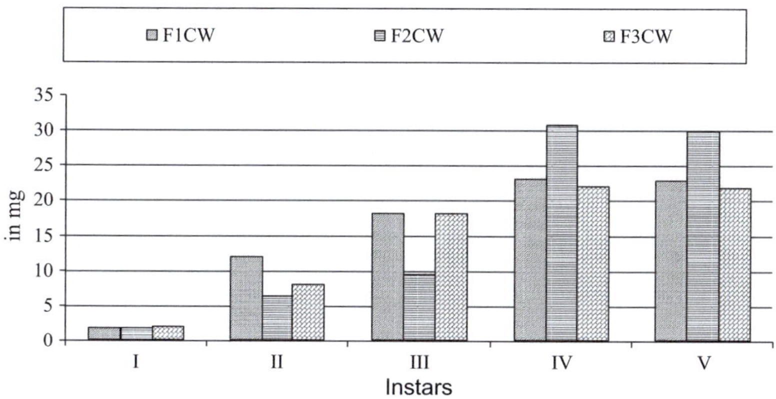

Fig. 2.5 Weight gain (mg) of *Rhynocoris marginatus* of different life stages reared on water once a week

predator/day). In this experiment *Rhynocoris marginatus* nymphs and adult (except second instar larvae and adult female) have the capacity to consume more number of *Spodoptera litura* larvae (76.1 mg/predator), when the predator was reared on oligidic diet (Table 2.9). Similarly insect hosts were reared by *Rhynocoris marginatus* second and fifth nymphal instars and adult consumed by more number of *Spodoptera litura* and other life stages consumed *Dysdercus cingulatus* second instar. But either insect host or artificial diet-reared *Rhynocoris marginatus* of different life stages consumed minimum number of *Corcyra cephalonica*. Hence this reduviid can be used as a biological control agent in crop where *Spodoptera litura* and *Dysdercus cingulatus* are present.

Weight gain was also recorded for predator reared with artificial diet (Fig. 2.4) and *Corcyra cephalonica* larvae and water. Results reveal that provision of artificial diet or water is important to maintain the development and reproduction of this reduviid. Hence, weight gain gradually increased while the predator grew older (Fig. 2.5).

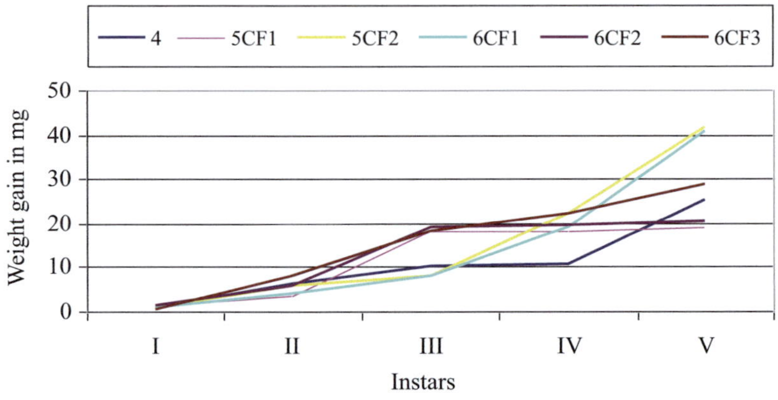

Fig. 2.6 Weight gain (in mg) of *Rhynocoris marginatus* reared on artificial diet with *Corcyra cephalonica* once a week

Instead of water, reduviid was reared with artificial diet and was also provided with the laboratory host *Corcyra cephalonica* larvae once a week for three consecutive generations, and their weight gain was recorded (Fig. 2.6). Predator gained weight and indicates the development of support to artificial diet. However, the weight gain slightly decreased in the second generation and subsequently in the third generation.

2.7 Conclusions

Hereditary feeding acts of *Rhynocoris marginatus* could not be changed in the artificial diet which shows there are lots of scopes for utilizing these diets for mass rearing of reduviid bugs. Results also reveal that the liquid diet should be provided using cotton as a medium for reduviid predator feeding. The recorded feeding acts were almost same as the reduviid feeding on insect as food. Information on food (artificial diet) acceptability can be obtained by recording in the laboratory. However, this can be done with caution, because, of the artificial condition, the predator may not truly express its atrophic relationships.

2.8 Future Recommendations

- Chemical composition of various natural as well as laboratory hosts will be studied, and the database can be maintained in order to develop an appropriate artificial diet for reduviids.
- Primary nutrients of artificial diet should be analyzed and compared with natural as well as laboratory hosts. Based upon the analysis, a new diet can be developed in the near future.

- Minerals and amino acid compositions should be studied in the near future.
- Identification of info-chemicals which attract reduviid for feeding the diet is needed.

References

Allaluya Jasmine C, Shanmuga S, Kombiah P, Kalidas S, Sahayaraj K (2012) Biosafety evaluation of *Tephrosia purpurae* stem-based formulation (Telp 3% EC) against three *Rhynocoris* spp. Asian J Biol Sci 5(4):216–220

Ambrose DP (1999) Assassin bugs. Science Publishers, Enfield

Ambrose DP, Saju T, Sahayaraj K (1990) Prey influence on the development, reproduction and size of the assassin bug *Rhynocoris marginatus*. Ibid. 8(1): 280–287

Ambrose DP, Kumar SP, Nagarajan K, Das SSM, Ravichandran B (2006) Redescription, biology, lifetable, behaviour and ecotypism of *Sphedanolestes minusculus* Bergroth (Hemiptera : Reduviidae). Entomologia Groatia 1–2:47–66

Avila TO, Woods HA, Raguso RA (2003) Effects of dietary variation on growth, composition, and maturation of *Manduca sexta* (Sphingidae: Lepidoptera). J Insect Physiol 49:293–306

Bonte M, De Clercq P (2008) Developmental and reproductive fitness of *Orius laevigatus* (Hemiptera: Anthocoridae) reared on factitious and artificial diets. J Econ Entomol 101:1127–1133

Bonte M, De Clercq P (2010) Impact of artificial rearing systems on the developmental and reproductive fitness of the predatory bug, *Orius laevigatus*. J Insect Sci 10(1):104

Castane C, Zapata R (2005) Rearing the predatory bug *Macrolophus caliginosus* on a meat-based diet. Biol Control 34:66–72

Chapman RF (1998) The insects: structure and function, 4th edn. Cambridge university press, New York

Cohen AC (1998) Solid-to-liquid feeding: the inside(s) story of extra-oral digestion in predaceous Arthropoda. Am Entomol 44:103–117

Cohen AC (2004) Insect diets science and technology. CRC Press, New York, p 324

De Clercq P (2004) Culture of natural enemies on factitious foods and artificial diets. In: Capinera JL (ed) Encyclopedia of Entomology, vol I. Kluwer Academic Publishers, Dordrecht, pp 650–652

De Clercq P, Degheele D (1992) A meat-based diet for rearing the predatory stinkbugs *Podisus maculiventris* and *Podisus sagitta* [Hetroptera : Pentatomidae]. Entomophaga 37(1):149–157

Edney EB (1977) Water balance in land arthropods. Springer-Verlag, Berlin

Grenier S, De Clercq P (2003) Comparison of artificially versus naturally reared natural enemies and their potential for use in biological control. In: van Lenteren J (ed) Quality control and production of biological control agents: theory and testing procedures. CABI Publishing, Wallingford, pp 115–131

Grenier S, Greany PD, Cohen AC (1994) Potential for mass release of insect parasitoids and predators through development of artificial culture techniques. In: Rosen D, Bennett FD, Capinera JL (eds) Pest management in the subtropics: biological control – a Florida perspective. Intercept, Andover, pp 181–205

Grenier S, Guillaud J, Delobel B, Bonnot G (1989) Nutrition et elevage du predateur polyphage *Macrolophus caliginosus* (Heteroptera, Miridae) sur milieux artificiels. Entomophaga 34:77–86

Guillermo-Ferreira R, Cardoso-Leite R, Gandolfo R (2012) First observation of alternative food usage (extrafloral nectar) by the assassin bug *Atopozelus opsimus* (Hemiptera: Reduviidae). Rev Bras Entomologia 56(4):489–491

Hagler JR, Naranjo SE (2005) Use of a gut content ELISA to detect whitefly predator feeding activity after field exposure to different insecticide treatments. Biocontrol Sci Tech 15:321–339

Haviland MD (1931) The Reduviidae of Kartabo, Bartica. District, British Guiana. Zoologica 7:129–154

Hill CJ (1989) The effect of adult diet on the biological of butterflies (The common crow butterfly). Oecologia 81(2):258–266

Holling CS (1966) The functional response of invertebrate predators of prey density. Entomol Soc Can 48:1–86

Irudayaraj V, Martin P, Selvaraj P, Sahayaraj K (2003) Predatory behaviour of *Rhynocoris marginatus* on *Danass chrysippus* Linn. Insect Environ 8(4):146

Karowe DN, Martin MM (1989) The effects of quantity and quality of diet nitrogen on the growth, efficiency of food utilization, nitrogen budget, and metabolic rate of fifth-instar *Spodoptera eridania* larvae (Lepidoptera: Noctuidae). J Insect Physiol 35:699–708

Kubota T, Shiga M (1995) Successive mass rearing of Chrysopids (Neuroptera: Chrysopidae) on eggs of *Tribolium castaneum* (Coleoptera: Tenebrionidae). Jpn J Appl Entomol Zool 39(1):51–58

Kumar SM, Sahayaraj K (2012) Gross morphology and histology of head and salivary apparatus of the predatory bug, *Rhynocoris marginatus*. J Insect Sci 11:154

Martin MM, Vant Hof HM (1988) The cause of reduced growth of *Manduca sexta* larvae on a low water diet: increased metabolic processing cost or nutrient limitation. J Insect Physiol 34:515–525

McEwen PK, Kidd NA (1995) The effects of different components of an artificial food on adult green lacewing *Chrysoperla carnea* fecundity and longevity. Entomol Exp Appl 77(3):343–346

Nagarajan K, Ambrose DP (2013) *Rhynocoris fuscipes* (Fabricius) (Insecta: Heteroptera: Reduviidae) by Y-arm olfactometer. Pak J Biol Sci 16(21):1363–1367

Narbona E, Dirzo R (2010) A reassessment of the function of floral nectar in *Croton suberosus* (Euphorbiaceae): a reward for plant defenders and pollinators. Am J Bot 97:672–679

Norioka N, Okada T, Hamazume Y, Mega T, Ikenaka T (1984) Comparison of nutritive value of egg yolk and egg white and whole egg. J Biochem 97:19–28

Ralston JS (1977) Egg guarding by male assassin bugs of the genus *Zelus* (Hemiptera: Reduviidae). Psyche 87:103–107

Sahayaraj K (2000) Evaluation of biological control potential evaluation of *Rhynocoris marginatus* (Fab.) on four groundnut pests under laboratory condition. Int Arachis Newsl 20(1):72–74

Sahayaraj K (2008) Approaching behaviour of *Rhynocoris marginatus* (Fab.) (Heteroptera: Reduviidae) on three prey kairomones. Bull Insectol 61(2):233–237

Sahayaraj K (2011) Hunter reduviids in pest management for plantation crop. In Proceeding of National Seminar on Harmful/Beneficial Insects of Agricultural Importance, Calicut, Kerala, India, p 42–51

Sahayaraj K, Balasubramanian R (2008) Biological control potential evaluation of artificial and factitious diets reared *Rhynocoris marginatus* (Fab.) on three pest. Arch Phytopathol Plant Protect 42(3):238–247

Sahayaraj, Karthikraja S (2003) Effect of biopesticides on *Rhynocoris marginatus* (Fab). J Biol Control 17(1):43–45

Sahayaraj K, Paulraj MG (1999a) Effect of plant products on the eggs of *Rhynocoris marginatus* (Fab.) (Hemiptera: Reduviidae). Insect Environ 5(1):23–24

Sahayaraj K, Paulraj MG (1999b) Toxicity of some plant extracts against the life stages of a reduviid predator *Rhynocoris marginatus* (Fab.). Indian J Entomol 61(4):342–344

Sahayaraj K, Paulraj MG (2001a) Effect of cold storage on egg hatching in two reduviid predators, *Rhynocoris marginatus* Fab. and *Rhynocoris fuscipes* Fab. Belgium J Entomol 3:201–207

Sahayaraj K, Paulraj MG (2001b) Rearing and life table of reduviid predator *Rhynocoris marginatus* Fab. (Heteroptera : Reduviidae) on *Spodoptera litura* Fab. (Lepidoptera : Noctuidae) larvae. J Appl Entomol 125(6):321–325

Sahayaraj K, Subasini M, Ravi C (2002) Influence of biorational insecticide *Coleus ambonicus* Lour (= *aromaticus* Benth) to *Spodoptera litura* (Fab.) (Lepidoptera: Noctuidae) and its predator *Rhynocoris marginatus* (Fab.) (Hemiptera: Reduviidae) in laboratory conditions. J Nat Conserv 14(1):113–122

Sahayaraj K, Martin P, Raju G (2003) Effect of temperature and water consumption of *Rhynocoris marginatus* Fab. (Hemiptera: Reduviidae). Entomon 28(2):175–177

Sahayaraj K, Thangarani S, Delma JCR (2004) Comparative prey suitability of *Helicoverpa armigera* and *Spodoptera litura* larvae for *Rhynocoris marginatus* (FAB.) (Heteroptera: Reduviidae). Belgium J Entomol 6:383–392

Sahayaraj K, Martin P, Selvaraj P, Raju G (2006) Feeding behaviour of Reduviid predators on meat and insect – based artificial diets. Belgean J Entomol 8:55–65

Sahayaraj K, Venkatesh P, Balasubramanian R (2007) Feeding behaviour and biology of a reduviid predator *Rhynocoris marginatus* (Fabricius) (Heteroptera: Reduviidae) on Oligidic Diet. Hexapoda 14(1):24–30

Sahayaraj K, Borgio JAF, Kumar SM (2012) First record of *Aspergillus flavus* as a fungal pathogen of the predator *Rhynocoris marginatus* (Hemiptera: Reduviidae). Entomol Brasilis J 5(1):80–81

Sahayaraj K, Kuumar V, Avery P (2015) Functional response of *Rhynocoris kumarii* (Heteroptera: Reduviidae) on *Phenacoccus solenopsis* (Hemiptera: Pseudococcidae) in the laboratory. Eur J Entomol 112(1):69–74

Slansky F (1993) Nutritional ecology: the fundamental quest for nutrients. In: Stamp NE, Casey TM (eds) Caterpillars: ecological and evolutionary constraints on foraging. Chapman and Hall, New York, pp 29–91

Stoner A, Metcalfe AM, Weeks RE (1975) Plant feeding by reduviidae, a predaceous family (Hemiptera). J Kansas Entomol Soc 48:185–188

Sujatha S, Sahayaraj K (2007) Influence of oligidic diet and factitious host on the development, survival and adult longevity of *Rhynocoris marginatus* (Fabricious) (Heteroptera: Reduviidae: Harpactorinae). In: Narayann M, Sethuramalingam T, Sahayaraj K (eds) National Seminar on Technology and Management of Bioresearches, p 35–39

Tallamy DW, Walsh E, Peck DC (2004) Revisiting paternal care in the assassin bug, *Atopozelus pallens* (Heteroptera: Reduviidae). J Insect Behav 17:431–436

Tan XL, Zhao J, Wang S, Zhang F (2015) Optimization and evaluation of microencapsulated artificial diet for mass rearing the predatory ladybird *Propylea japonica* (Coleoptera: Coccinellidae). Insect Sci 22(1):111–120

Taylor JR, Schmidt JM (1996) Factors regulating predation by first-instar spined assassin bugs (*Sinea diadema* (Fab.)) (Hemiptera: Reduviidae). J Insect Behav 9(1):23–35

Tebayashi SI, Kawahara T, Kim CS, Nishi A, Takahashi K, Miyanoshita A, Horiike M (2003) Feeding stimulants eliciting the probing behavior for *Peregrinator biannulipes* Montrouzier et Signore (Hemiptera: Ruduviidae) from *Tribolium confu*sum (Jacquelin du Val). Z Naturforsch C 58(3–4):295–299

Thakur A, Dhammi P, Saini HS, Kaur S (2016) Effect of antibiotic on survival and development of *Spodoptera litura* (Lepidoptera: Noctuidae) and its gut microbial diversity. Bull Entomol Res 106(3):387–394

Thompson SN, Hagen KS (1999) Nutrition of entomophagous insects and other arthropods. In: Bellows TS, Fisher TW (eds) Biological control: principles and applications of biological control. Academic, San Diego, pp 594–652

Timmins WA, Bellward K, Stamp AJ, Reynolds SE (1988) Food intake, conversion efficiency, and feeding behaviour of tobacco hornworm caterpillars given artificial diet of varying nutrient and water content. Physiol Entomol 13:303–314

Vandekerkhove B, Van Baal E, Bolckmans K, De Clercq P (2006) Effect of diet and mating status on ovarian development and oviposition in the polyphagous predator *Macrolophus aliginosus* (Heteroptera: Miridae). Biol Control 39:532–538

Xie ZN, Wu ZX, Nettles WC, Saldang A, Andnordlundd A (1997) In vitro culture of *Trichogramma* spp. on artificial diets containing yeast extract and ultracentrifuged chicken egg yolk but devoid of insect components. Biol Control 8:107–110

Biology

3

Abstract

Six artificial diets, devoid of insect components, were developed based on the feeding preference test for the predator *Rhynocoris marginatus* for mass production to control insect pest in agricultural ecosystem. *Rhynocoris marginatus* nymphs were successfully reared to adult in meat-based artificial diet, and these adults were subsequently able to reproduce. These results indicated that it is possible to rear this predator on a diet that completely excludes insect material. The diet based on pork liver was used as major source to approach the nutritional characteristics of the primary insect prey, *Spodoptera litura. Rhynocoris marginatus* was reared on different diets for three consecutive generations. Developmental time and preovipositional period were significantly longer, and egg viability, survival from egg to adult, and fecundity were significantly lower in *Rhynocoris marginatus* individuals reared on artificial diets than in those reared on *Spodoptera litura*. A significant reduction in developmental time, increase in survival from egg to adult, and reduction in length of the preovipositional period were observed after 3 generations of artificial diet-reared *Rhynocoris marginatus* on diet six. These changes may indicate that the predators experience some degree of adaptation to the diet after several consecutive generations of artificial diet rearing. No changes in egg viability were observed after three generations of artificial diet rearing. Although the average fecundity of a female was low for the oligidic diet was sufficient for sustaining continuous generation of *Rhynocoris marginatus*. There are three generations that have been obtained through continuous culture of this reduviid-fed meat diet. The meat-based diet tested was able to sustain the rearing of *Rhynocoris marginatus* for several generations without supplying any insect pest showing a good nymph survival rate. This study clearly shows that *Rhynocoris marginatus* can be reared on an oligidic diet that is economically producible and the fecundity of the female is significantly less than those reared on either *Corcyra cephalonica* or *Spodoptera litura.*

K. Sahayaraj, R. Balasubramanian, *Artificial Rearing of Reduviid Predators for Pest Management*, DOI 10.1007/978-981-10-2522-8_3

Keywords
Predators rearing • Biological traits • Development • Survival • Fecundity • Weight gain

3.1 Meridic Diet for Rearing Hemipteran Predators

3.1.1 Necessity of Predatory Bug's Mass Production

The greatest barrier for the mass production of predatory insects is the lack of suitable artificial diets. Singh (1977) noted the 754 species of arthropod had been reared on artificial diets, of these 27 were arachnids. The remaining are insect species spanning 10 orders consisting of 19 families of Coleoptera, 24 Diptera, 11 Homoptera, 8 Hymenoptera, and 27 Lepidoptera. Sikorowski and Goodwin (1985) also reported similar numbers. However, Waage et al. (1985) pointed out that no suitable artificial diets had been developed for predators. An oligidic diet free of insect components was developed and sustained (says Cohen 1985a, b). The status of in -vitro culture of parasitic insect has been reviewed by House (1967) and Thompson and Hagen (1999). In spite of some promising results obtained in the development of artificial diet for entomophage, the use of artificial diets in mass propagation programs is currently limited to only a few species of predators and parasitoids (Slansky and Rodriguez 1987; Waage et al. 1985; Tan et al. 2015).

3.1.2 Artificial Diets for Hemipteran Predator Rearing

Artificial diet prepared by using meat, insect as source ingredients, has been utilized for rearing the predators of Pentatomidae, Anthocoridae, Miridae, and Lygaeidae. However, very limited literature is available for reduviid predators. All available information was carried out by the author only.

3.1.2.1 Pentatomidae

In the beginning, a diet based on bovine meat and was developed and reared on both *Podisus maculiventris* (Say) and *Podisus sagitta* (Fabricius) for seven and five continuous generations, respectively, in comparison to rearing on larvae of the greater wax moth, *Galleria mellonella* (L.). Results reveals that nymphal development was prolonged with 15–40 % and adult weights were lower, reaching 72–82 % of the control weights. Fecundity of females reared on the meat diet was reduced to about 1/3–1/2 of the control (De Clercq and Degheele 1992). During the subsequent year, these predators were reared for more than 15 consecutive generations on a meat-based artificial diet. When they were returned to a diet of live prey (larvae of the pyralid *Galleria mellonella*) after different generations on the artificial diet, developmental and reproductive traits were similar to bugs continuously reared on live prey (De Clercq and Degheele 1993).

Development and reproduction of the predatory pentatomid *Podisus maculiventris* (Say) was assessed on three types of unnatural prey and two presentation systems of an artificial diet. Predator nymphs developed faster on pupae of the yellow mealworm, *Tenebrio molitor* L., than on larvae of the beetle or caterpillars of the greater wax moth, *Galleria mellonella* L., but nymphal survival and adult weights were similar on all preys tested. Females reared on wax moth larvae, rarely, produced twice as many eggs (i.e., ca. 1000 eggs per female) as those fed with larvae or pupae of the tenebrionid.

Overall performance of *Podisus maculiventris* on two presentation methods of an oligidic artificial diet based on bovine meat was inferior to live prey. Packaging the meat diet in stretched Parafilm sheets yielded better results than a gelled form of the diet. Female predators reared on the artificial diets were significantly smaller than those produced on live prey, but had a fecundity similar to females fed with on larvae or pupae of *T. molitor.* The practical value of the different foods for the mass rearing of *Panthous maculiventris* is discussed (De Clercq et al. 1998). For the development of *Brontocoris tabidus,* an artificial diet based on beef meat and liver was evaluated by Zanuncio et al. (1996). The predator significantly showed longer nymphal development and lower adult weights on this diet than when reared on pupae of the mealworm *Tenebrio molitor.*

An artificial diet comprised of pig liver and tuna fish but devoid of insect components was developed for continuous rearing of the predator *Arma chinensis* Fallou. Our results demonstrated less desirable biological parameters in *A. chinensis* reared on this artificial diet compared to a secondary prey, pupae of Chinese oak silk moth *Antheraea pernyi (*Guérin-Méneville). Weight of eggs and adults, fecundity, and egg viability were lower for diet-fed *Antheraea chinensis* compared to *Antheraea chinensis* reared on pupae of *Antheraea pernyi.* Developmental time from second instar to adult and the preovipositional period were significantly longer for diet-fed *Antheraea chinensis.* Cannibalism was also higher with diet-fed *Antheraea chinensis.*

Over successive generations of rearing on the diet, some of the characteristics improved, such as developmental time, survival from second nymphal instar to adult, sex ratio (♀/♂:♀), adult weight, fecundity, and fertility. These changes may indicate that the predators experience some degree of adaptation to genetic selection for the diet after several consecutive generations. There were no changes in developmental time of egg and first instar or survival from first to second instar with successive generations reared on the artificial diet. However, diet-fed adults did live longer than pupa-fed adults.

Rojas et al. (2000) tested two meridic diets on the biological traits of *Panthous bioculatus*. Results show that developmental time and preovipositional period were longer and egg viability, survival from egg to adult, and fecundity were lower in *Panthous bioculatus* individuals reared on artificial diets than in those reared on *L. decemlineata* eggs. Coudron and Kim (2004) showed that when fed with an insect-free zoophytogenous artificial diet during both the nymphal and adult stages, development times of *Panthous bioculatus* were prolonged, and the net reproductive rates and the intrinsic rates of increase were significantly lower than when fed with larval *T. ni* at both nymphal and adult stages.

3.1.2.2 Anthocoridae

Initially in *Orius insidiosus* (Say) and others, artificial medium made from carrageenan salt of potassium chloride and covered with paraffin wax to rear Orius spp is proposed (Castane and Zalom 1994). Further, the liver-based artificial diets were developed for the production of *Orius laevigatus* Fieber (Arijs and De Clercq 2001, 2002, 2004). Results show that nymphal development was slower, on artificial diets than on the control food [*Ephestia kuehniella* Zeller (Pyralidae)], taking 15.0–15.9 days versus 14.3 days, respectively. Further results suggested that artificial diets containing liver and egg yolk as the main components may prove useful to make mass production of *Orius laevigatus* and other heteropteran predators more cost effective. *Orius insidiosus* (Say) was also reared on an artificial diet absent of insect components, but yolk production is reduced and resultant rates of oviposition and egg production are poor. Supplementing the artificial diet with cells from an embryonic cell line, IPLB-(PiE) of *Plodia interpunctella* (Hübner), it enhanced oviposition rates (Ferkovich and Shapiro 2004a, b).

Latter Arijs and De Clercq (2004) used different meat- and liver-based artificial diets where were designed for the generalist predator *Orius laevigatus* (Fieber) (Anthocoridae), and their suitability was compared to that of its factitious food, *Ephestia kuehniella* Zeller (Pyralidae) eggs. They suggested that artificial diets containing liver and egg yolk as the main components may prove useful to make mass production of *Orius laevigatus* and other heteropteran predators more cost effective.

3.1.2.3 Lygaeidae

Geocoris punctipes Say was reared for more than 90 generations on a meridic diet composed mainly of ground beef and beef liver (Cohen 1985; Cohen and Staten 1994; Cohen and Urias 1986a, b). A simple diet based on beef and liver was successfully developed for *Geocoris punctipes* Say (Lygaeidae) (Cohen 1985a; Cohen and Smith 1998).

3.1.2.4 Miridae

Predatory miridae, these diets have been successfully used for rearing *Dicyphus tamaninii* Wagner, a Mediterranean species (Iriarte and Castane 2001) reared *Dicyphus tamaninii* on the diet described for *Panthous maculiventris* (De Clercq and Degheele 1992). Firlej and co-workers (2006) compared biological parameters of the mite predator *Hyaliodes vitripennis* reared on live *Tetranychus urticae* Koch versus two artificial diets. These results demonstrate that *H. vitripennis* could be reared on an artificial diet and suggest that β-sitosterol has phagostimulant properties for this predator.

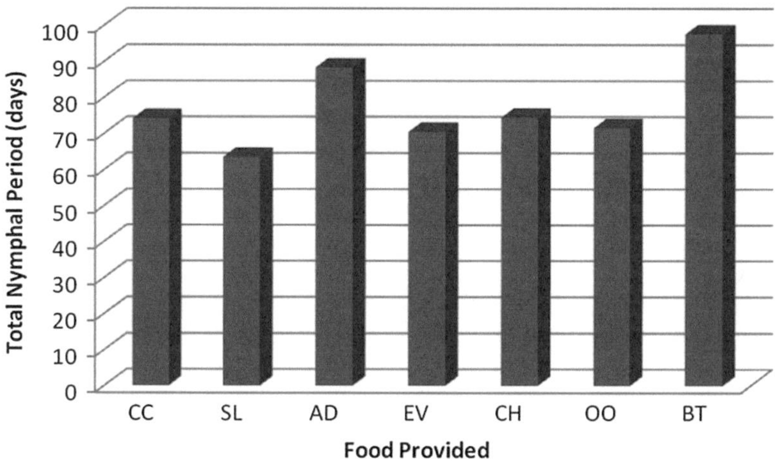

Fig. 3.1 Total nymphal developmental period (days) of *Rhynocoris kumarii* reared on *Corcyra cephalonica* (CC), *Spodoptera litura* (SL), artificial diet (AD), *Chrotogonus* sp. (CH), *Odoentotermes obesus* (OO), and *Earias vitella* (EV)

3.2 Meridic Diet on Chosen Reduviidae Biology

Incubation period is shortest in *Rhynocoris kumarii* (10 days) than *Sycanus collaris* (15 days) and *Panthous bmaculatus* (21 days). Sundararaju (1984) reported that *S. collaris* incubation period was 11 days (Sahayaraj 2012). Further he recorded that the total nymphal developmental period was 75.67 ± 9.06, 88.30 ± 3.60, and 101.12 ± 2.30 for *Sycanus collaris, Rhynocoris kumarii,* and *Panthous bmaculatus.* The total nymphal duration of *Rhynocoris kumarii* was 49.3 ± 1.95 days (George et al. 1998). It was extended more than 1.7 times while it was reared with artificial diet. This indicates that artificial diet composition could be modified for the shortening of the nymphal developmental time of the reduviids. However, the total nymphal developmental period of *Rhynocoris kumarii* varied between 63 and 98 days depending on the type of prey offered (Sahayaraj 2007; Sahayaraj et al. 2007a) (Fig. 3.1). Hence, the proposed diet can also be used for the rearing of *Rhynocoris kumarii* too. The total nymphal survival rate was higher for *Rhynocoris kumarii* (62.94) than for *Sycanus collaris* (46.16) and *Panthous bmaculatus* (16.23). The total nymphal survival rate was higher for *Rhynocoris kumarii* (62.94) than for *Sycanus collaris* (46.16) and *Panthous bimaculatus* (16.23).

3.3 Meridic Diet on *Rhyncoris marginatus* Biology

Initially, artificial diet was checked against the feeding behavior of *Rhynocoris marginatus* Fab. (Sahayaraj et al. 2006). *Rhynocoris marginatus* (Sahayaraj et al. 2006; 2007a, b; Sahayaraj 2008) has been reared with oligidic diets. For the first time, rearing of three reduviid predators like *Sycanus collaris, Panthous bmaculatus,* and

Panthous bmaculatus using meat-based artificial diet under laboratory conditions has been introduced (Sahayaraj 2012). Incubation period and total nymphal developmental period of *Rhynocoris kumarii* Ambrose and Livingstone, *Sycanus collaris* and *Panthous bmaculatus* (21 days) suggest that these reduviids can be reared using meat-based artificial diets and can be utilized in pest management program.

3.3.1 Biological Traits

A laboratory colony of *Rhynocoris marginatus* was maintained with individual collection from the agricultural and nearby scrub jungle ecosystems of Tirunelveli District, Tamil Nadu, India. The stock colony were cultured under laboratory conditions (30 ± 1 °C and 80 % RH and 11L:13D Hrs) on *Corcyra cephalonica,* as well as natural prey *Spodoptera litura. Spodoptera litura* was collected from cotton agroecosystems of Tamil Nadu, India, and mass reared on their natural host. Laboratory emerged *Rhynocoris marginatus,* and pests were used for this experiment. *Corcyra cephalonica* was also maintained under same laboratory conditions using the methodology of Sahayaraj (2002).

Nymphal development of *Rhynocoris marginatus* was monitored on five diets: fourth instar larvae of *Corcyra cephalonica* (T_1), and *Corcyra cephalonica* weekly once with water (T_2), oligidic diets (OD), (4, 5, and 6) (T_3), OD weekly once with *Corcyra cephalonica* (T_4), and *Spodoptera litura* alone (T_5). For each treatment, 100 newly hatched first instar *Rhynocoris marginatus* were randomly taken from laboratory culture and placed individually in plastic vials (6 cm height and 4.5 cm diameter) Cotton soaked in. water (5 mg) and oligidic diet were provided to the reduviids. For T1-T5 categories for nymphs (except T3), reduviids were also provided with one or two preys per day. Fresh weight of *Spodoptera litura* and *Corcyra cephalonica* larvae was 200 mg and 150 mg, respectively. Larvae of both prey species were made partly defenseless by crushing their head capsules; this also prevents the larvae from web spinning. Biological parameters like nymphal developmental periods, weight, survival rate, and sex ratio (♀/♀/♂) of *Rhynocoris marginatus* were recorded upon emergence for each category. The nymphs and adults were weighed on a Dhona, monopan balance (±0.1 mg). Occurrence of morphological abnormalities if any was also recorded.

3.3.2 Nymphal Development

In order to develop a rearing program, it is necessary to study the biology of the beneficial species and insect host and to seek information on plant hosts or artificial diets. Reduviid predator nymphs were successfully reared to adult on meat-based diet, and these adults were subsequently able to reproduce. This result indicates that it is possible to rear this predator on a diet which completely excludes insect material. Although the average fecundity of a female was low for the oligidic diets, nymphal survival rate and longevity of female were good. The oligidic diet was

Table 3.1 Nymphal developmental period (in days) of *Rhynocoris marginatus* reared with oligidic diets

	Developmental stages (days)					
Diet	I	II	III	IV	V	Total
3	9.2 ± 0.5	16.6 ± 1.0	15.9 ± 1.3	–	–	–
4	12.8 ± 0.6	17.6 ± 1.0	18.1 ± 1.3	17.3 ± 1.1	16.5 ± 0.9	80.7
5F1	13.0 ± 0.9	15.7 ± 1.8	20.2 ± 1.8	12.2 ± 3.0	17.8 ± 1.7	75.1
5F2	9.5 ± 0.3	12.2 ± 0.5	16.8 ± 0.9	13.0 ± 0.8	20.1 ± 1.1	71.9
6F1	6.6 ± 0.2	13.9 ± 0.9	12.9 ± 0.9	14.8 ± 0.7	21.8 ± 0.7	70.1
6F2	7.8 ± 0.4	7.1 ± 0.4	13.8 ± 0.9	15.7 ± 1.9	24.2 ± 1.5	68.8
6F3	8.1 ± 0.3	12.1 ± 0.6	12.9 ± 0.5	13.1 ± 0.6	18.8 ± 1.3	65.1

F1, F2, and F3 stands for first, second, and third filial generations, respectively; means ± means within a column and between instars followed by the same letter are not significant; $p > 0.05$; "*t*" test

sufficient for sustain continuous generation of *Rhynocoris marginatus*. Three generations were obtained through continuous cultures of this reduviid fed with meat diet which shows that the predator accepts the food. Information on food (artificial diet) acceptability can be obtained by recording in the laboratory. However, this can be done with caution, because of the artificial conditions; the predator may not truly express its tropic relationships. The meat-based diet tested was able to sustain the rearing of *Rhynocoris marginatus* for several generations without supplying any insect material, showing a good nymph survival rate.

During 2002, a total of 210 *Rhynocoris marginatus* were provided with diet 3, and the developmental time of nymphs was recorded. Fifth stadium, fourth stadium, third stadium, second stadium, and first stadium took 52.5, 35.3, 20.7, 16.2, and 23.3 days, respectively. In total, a nymph needs 147.8 days to attain in to adults. In 2003, Sahayaraj and Sujatha (unpublished data) reared *Rhynocoris marginatus* using beef liver-based artificial diet (with 30g/L and 60g/L). *Rhynocoris marginatus* took 123 and 143.6 days, respectively to complete its nymphal period. But the predator does not lay any fertile eggs.

Rhynocoris marginatus was fed with five different diets such as *Corcyra cephalonica* (T1), *Corcyra cephalonica* weekly once with water (T2), oligidic diet (T3), oligidic diet and weekly once *Corcyra cephalonica* (T4), and *Spodoptera litura* (T5). Previously, in 2003, Sahayaraj and co-workers recorded water consumption of *Rhynocoris marginatus* under laboratory conditions, hence we provided water weekly once to this predator. The nymphal developmental period of *Rhynocoris marginatus* reared on artificial diet; natural host, *Spodoptera litura;* and factitious host, *Corcyra cephalonica* is presented in Tables 3.1–3.3. Results revealed that rearing of *Rhynocoris marginatus* with oligidic diet prolonged nymphal developmental period. For instance, the total nymphal developmental period of *Rhynocoris marginatus* on artificial diet ranged from 65–81 days (Table 3.1) when compared with 45–47 days for *Corcyra cephalonica* (Table 3.2).

Initially *Rhynocoris marginatus* was reared on oligidic diets such as diets 1, 2, and 3 as different source ingredients. In these diets *Rhynocoris marginatus* did not

Table 3.2 Nymphal developmental period (in days) of *Rhynocoris marginatus* reared on *Spodoptera litura* (T_5) *Corcyra cephalonica* alone (C) (T_1) and *Corcyra cephalonica* once with water (CW) (T_2)

	Developmental stages					
Diet	I	II	III	IV	V	Total
Corcyra cephalonica						
F1	6.5 ± 0.1	8.6 ± 0.2	9.5 ± 0.6	9.8 ± 0.3	15.3 ± 0.6	46.8
F2	7.6 ± 3.1	7.1 ± 1.1	7.9 ± 2.4	7.8 ± 1.8	13.0 ± 0.2	43.9
F3	6.5 ± 0.1	8.0 ± 0.2	8.0 ± 0.6	8.2 ± 0.2	13.6 ± 0.5	45.2
Corcyra cephalonica + water						
F1	7.8 ± 0.4	11.0 ± 0.6	10.5 ± 0.8	11.3 ± 0.9	15.0 ± 0.6	55.7
F2	7.3 ± 0.1	9.9 ± 0.2	8.0 ± 0.1	11.4 ± 0.2	15.3 ± 0.3	51.9
F3	7.5 ± 0.1	9.7 ± 0.2	8.4 ± 0.2	10.7 ± 0.3	15.2 ± 0.4	51.6
Spodoptera litura						
F1	6.0 ± 0.2	6.95 ± 0.2	7.4 ± 0.2	6.4 ± 0.1	14.1 ± 0.2	41.1
F2	5.9 ± 0.1	6.2 ± 0.2	7.1 ± 0.1	7.1 ± 0.1	14.6 ± 0.1	41.1
F3	6.0 ± 0.0	6.8 ± 0.2	6.6 ± 0.5	8.8 ± 0.9	14.4 ± 1.2	42.6

F1, F2, and F3 stands for first, second, and third filial generations, respectively; means ± means within a column and between instars followed by the same letter are not significant; $p > 0.05$; "*t*" test

proceed from first instar to third instar even after 50, 40, and 30 days, respectively. Similarly diets 4 and 5 did not show satisfactory results. Hence their composition was modified and diet 6. was prepared. In diet 5 ingredients such as pork blood, vitamins E and C, and then acetic acid were added. In this diet, the total nymphal developmental period was longer (80.7 days). Hence, we included blood serum (2 ml) and considered it as diet 5. Addition of blood serum reduced *Rhynocoris marginatus* nymphal developmental period which was 75.1 days. In diet 6, multivitamin content was increased from 200 to 300 mg. It was further reduced to 70.1 days in diet (Table 3.1) when *Rhynocoris marginatus* was reared on the same diet; the total nymphal developmental period was reduced to 68.8 and 65.1 days in F2 and F3 filial generations.

Shortest (6.6 days for 6F1G) and longest (24.2 days for 6F2G) nymphal developmental periods were observed in the first and fifth instars, respectively ($t = -7.584$, df = 25, $P = 0.0005$). In T2 category total nymphal period ranged from (CWF1) 55.7–51.6 days (F3G) ($t = 7.726$, df = 28, $P = 0.0005$). In control category, total developmental period was ranged from 46.8 to 43.9 days (CF2) (Table 3.2). Moreover, we observed maximum mortality in the third instar. Nymphal developmental period was too longer than control diet so this diet also not gives satisfactory results for the rearing of this bug.

But it was reduced to 63.6 days (average 64.73 days) (T4 F1), if *Corcyra cephalonica* was provided once in a week ($t = -6.947$, df = 2, $P < 0.020$). The total nymphal developmental period was further reduced to 55.7 days in F1 generation (average 53.06 days), when water is provided once in a week along with *Corcyra cephalonica* (Table 3.3). During the same experimental situation, *Rhynocoris marginatus*

Table 3.3 Nymphal developmental period (in days) of *Rhynocoris marginatus* reared on oligidic diets with *Corcyra cephalonica* (T_4)

	Developmental stages					
Diet	I	II	III	IV	V	Total
3C	8.4 ± 0.2	13.6 ± 0.7	16.3 ± 0.3	18.2 ± 3.4	–	–
4C	7.3 ± 0.4	14.4 ± 0.8	18.2 ± 1.1	12.8 ± 0.6	15.6 ± 0.9	68.8
5CF1	8.6 ± 0.5	11.6 ± 0.6	12.6 ± 0.8	12.0 ± 1.2	18.6 ± 1.2	63.6
5CF2	12.0 ± 0.3	13.8 ± 0.6	12.7 ± 0.7	11.5 ± 0.6	16.9 ± 0.9	66.9
6CF1	7.0 ± 0.1	10.2 ± 0.5	13.4 ± 0.9	11.3 ± 0.7	19.1 ± 1.1	61.1
6CF2	9.3 ± 0.3	13.0 ± 0.5	12.8 ± 0.6	13.6 ± 0.7	17.2 ± 1.0	64.3
6CF3	9.2 ± 0.2	13.5 ± 0.5	10.7 ± 1.1	12.8 ± 1.1	17.4 ± 1.2	63.7

F1, F2 and F3 stands for first, second, and third filial generations, respectively; Means ± means within a column and between a instar followed by the same letter are not significant $p > 0.05$; '*t*' test

developed within 43.9 days (average 45.3 days) in *Corcyra cephalonica* ($t = -3.694$, df = 2, $P < 0.066$) and natural diet 41.1 days (average 41.6) ($t = -16.288$, df = 2, $p < 0.004$). The total nymphal developmental period of *Rhyocoris marginatus* with *Spodoptera litura* (mean 41 days) and *Corcyra cephalonica* (mean, 42 days) was statistically significant. The direct way of nutrients can influence predator biology is not only nymphal development but also nymphal mortality. The results of this study indicate that when *Rhynocoris marginatus* is fed with either *Corcyra cephalonica* or *Spodoptera litura*, the predator attain adulthood and also reproduces.

Previously, it was reported that *Rhynocoris marginatus* completed nymphal development with in 46.71 days (Sahayaraj and Paulraj 2001) on *Spodoptera litura*; 84.70 days on *Odontotermes obesus* (Ambrose et al. 1994), and 45.0 days on *Helicoverpa armigera* (Sahayaraj et al. 2004). In OD *Rhynocoris marginatus* nymphal developmental period was 65 days. However, when it compares to the results of Ambrose et al. (1990), in OD *Rhynocoris marginatus* development reduces to 20 days. Moreover, in milk powder-based artificial diet, the total nymphal period of *Rhynocoris marginatus* was 147.8 days (Sahayaraj et al. 2007a, b) (Fig. 3.2).

Podisus maculiventris developed on meat diet slower than on wax moth larvae (De Clercq et al. 1998b). Although cannibalism occurred more frequently in reduviid reared with alive preys, the high yield of adults and eggs in consecutive generations of *Rhynocoris marginatus* does not support the conclusion that cannibalism may account for the success in development and reproduction. Furthermore, *Rhynocoris marginatus* was known to be with highly cannibalistic behavior when they were reared on their natural host.

3.3.3 Nymphal Survival Rate

In beef liver-based artificial diet diet 3 was used as a food for rearing 210 *Rhynocoris marginatus.* In total, only 11 (5.24 %) nymphs attained in to the adults. A greater percentage of nymphs survived in the first stadium (75 %), than in the fourth stadium (37.5 %), third stadium (36.6 %), fifth stadium (33.3 %), and second stadium

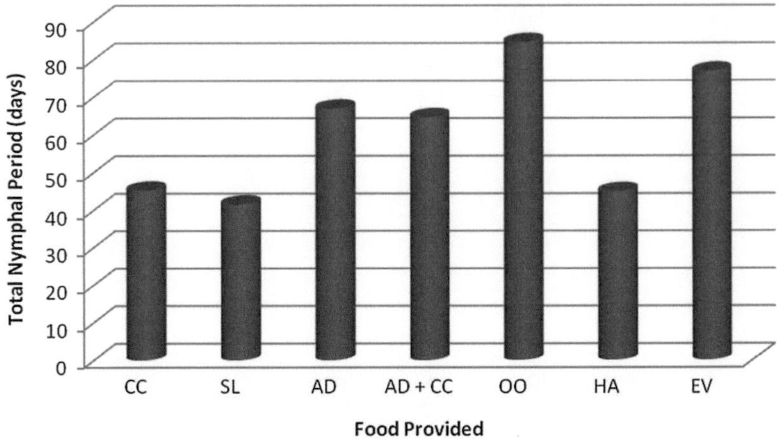

Fig. 3.2 Total nymphal developmental period (days) of *Rhynocoris marginatus* reared on *Corcyra cephalonica* (CC), *Spodoptera litura* (SL), artificial diet (AD), artificial diet + *Corcyra cephalonica* (CC), *Odontotermes obesus* (OO), *Helicoverpa armigera* (HA) and *Earias vitella* (EV)

(19 %). Sahayaraj and Sujatha (2012) reported that the survival rate of *Rhynocoris marginatus* was only 25 and 15.64 % when beef liver was included in the artificial diet at 30 g/L and 60 g/L, respectively.

In T_1 category, total nymphal survival rate was 92.17 (ranged from 80.56 % to 98.3 %). Nymphal survival rate was high in the first stadium (100 %) and minimum survival rate was observed in third generation (80.55 %). Among the three generations, CCF3 generation has the highest survival rate (98.26%) and lowest in F1generations the lowest, (80.55 %). In T2 category average nymphal survival rate was 92.17 %. Among the three generation F3 generation has the maximum survival rate (96.49 %) and minimum 85.71 %. As, observed in T1 category, among the nymphal instars the first instar has the maximum survival rate and second instar the minimum (70.00 %).

In T3 category, average survival rate was 80.91 %. Among the three generations, maximum survival rate was observed as 91.71 % and minimum survival rate was in 4A oligidic diet (T3) 37.50 %. Individual survival rate (100 %) was maximum in the first and second instar and minimum in the fourth instar (62.50 %). In T_4 category, the average nymphal survival rate was 85.73 %. Among the three generations maximum survival rate was observed in F3 generation and minimum in diet 4 of T4 generation (57.69 %). Individual maximum survival rate was observed in first and second instars (100 %) followed by minimum survival rate in fourth instar (72.72 %). Individual maximum survival rate was observed in first and second instars (100 %) followed by minimum survival rate in fourth instars (72.72 %).

In T_5 category, maximum survival rate was observed in the first stadium (100 %). Maximum survival rate observed in SLF2 generation (98.25 %). Average nymphal survival rate was maximum in (SL) T_5 category (94.55 %). In *Rhynocoris marginatus* adult female survival rate was low (75–76 days) when artificial diet was offered without insect material and there is no reproduction earlier in the diets. All of the diets had nutritional qualities that allowed complete development of the predators, to a greater or a lesser extent, indicating that there is a potential for rearing this insect on artificial media and that the artificial diet used here was adequate, allowing the nymph to acquire food and develop to the adult stage (De Clercq et al. 1998). Moreover, adult weight was also very low.

Our results are closer to those presented by Cohen (1995) and Cohen and Urias (1986a, b). The oviposition was not increased when *Corcyra cephalonica* larvae were provided along with the artificial diet. However, predatory rate was higher when compared *Rhynocoris marginatus* reared on with *Corcyra cephalonica* alone. It is possible that some minor imbalance in the nutrition of meat fed predators could generate a higher predation rate on *Corcyra cephalonica.* It should also be noted that the data analyzed in the present work relate to first generation meat reared predators and that in successive generation, life history traits could be improved due to adaptation to the diet as reflected by the increase in nymphal survivorship from the first (80.55 %) to the second (92.76 %) and third (98.26 %) generations (Wittmeyer and Coudron 2001).

3.3.4 Weight Gain

Initially it was observed that incorporation of beef liver in the artificial diet of *Rhynocoris marginatus* enhances the weight from 2.41 to 68.33 mg. It was further enhanced when 60 g beef was added in to the artificial diet of this predator (2.1–74.4 mg from first stadium to adult predator) (Sahayaraj and Sujatha 2012). Fresh weights of the newly emerged *Rhynocoris marginatus* were affected by diets during their nymphal periods. Among the experimental groups, the weight gain was higher in predators maintained on diet 5 than those fed with diet 4 (Fig. 3.1). However, the weight gain was significantly higher in the control category than *Spodoptera litura* (Table 3.4). In general, the weight gain was gradually increased from the first instar to the adults. But weight gain was reduced when water was provided along with *Corcyra cephalonica* in all the generations (CW T_2).

Meat diet produced smaller, as well as lighter-weight predator with longer developmental period , and weight gain was increased in embryonic and nymphal development times than the control categories (Fig. 3.2). In the experimental group, *Corcyra cephalonica* was provided along with AD, and weight gain of some of the instars was maximum for the predator maintained on T3 (5CF2, 21.66 mg and 16CF1, 20.92 mg) in among all the diets.

The weight gain was significantly higher in the control category. The weight gain of predators reared on the meat-based diet was about lower than control diet, where the average fresh weight of the predator was 30.92 and 38.83 mg, respectively. But in oligidic diet with *Corcyra cephalonica* category, weight gain of first, fourth. and fifth instars was also increased (1.24, 15.72 and 20.14 mg). The weight gain of the predators was significantly improved when supplied with *Corcyra cephalonica* larvae along with AD. Compensatory feeding was effective for making up small differences in diet nutrient content, but the lowest nutrient content diet was outside the

Table 3.4 Sex ratio and adult longevity (in days) of *Rhynocoris marginatus* reared with different preys

Prey	Adult longevity		
	Male	Female	Sex ratio
Spodoptera litura			
F1	86.9 ± 2.9	88.9 ± 2.9*	0.78
F2	82.2 ± 2.1	92.5 ± 1.2*	0.85
F3	71.3 ± 5.9	96.5 ± 2.3*	0.75
Corcyra cephalonica			
F1	109.8 ± 1.8	122.0 ± 3.9*	0.72
F2	102.0 ± 1.2	121.0 ± 0.1*	0.65
F3	108.0 ± 1.2	112.0 ± 0.2*	0.82
Corcyra cephalonica + water			
F1	96.4 ± 2.3	102.7 ± 5.0*	0.74
F2	55.8 ± 2.7	80.5 ± 4.6*	0.77
F3	65.1 ± 5.7	96.1 ± 8.4*	0.78

F1, F2 and F3 stands for first, second, and third filial generations, respectively; '*t*' test $p < 0.05$ % level
*indicates significance at 5 % level

range of compensatory ability. Growth rates are dependent on development time, the only major difference in growth rates between diet groups occurred on the lowest nutrient content of the diet. Weight gains of T_3 and T_4 categories were lower than other categories. A similar trend was a common feature in artificial diet reared insects (De Clercq and Degheele 1992; Freitas et al. 2006; Castane and Zapata 2005; Cohen 1985, 1990, 2000). This finding reinforces that compensatory feeding was effective for making up small differences in diet nutrient content, but the lowest nutrient content diet was outside the range of compensatory ability (Lepis and Travis 1994; Bradshaw and Johnson 1995; Nylin and Gotthard 1998; Flanagin et al. 2000). Though a direct correlation was observed between the adult weight and its fecundity, this finding did not support the statement. From the result it was very clear that the meat-based artificial diet produced smaller, lighter weight *Rhynocoris marginatus* with longer embryonic and nymphal development times than the predator reared on *Corcyra cephalonica* and *Spodoptera litura*; this is a commonly reported feature among artificially reared insects, as discussed in Grenier and De Clercq (2003).

3.3.5 Sex Ratio and Adult Longevity

Addition of beef liver in artificial diet altered the sex ratio of *Rhynocoris marginatus* (1.0: 0.5 ad 1.0: 0.57 for male and female at 30 and 60 g /L beef liver diet, respectively) (Sahayaraj and Sujatha 2012). Irrespective of the categories, the sex ratio was female biased and it was varied from 0.62 to 0.85 (Tables 3.4 and 3.5). The sex ratios of adults maintained in oligidic diets 5 and 6 were 0.62 and 0.76, respectively (chi square = 0.47, df = 5, $p = 0.99$), similarly sex ratio in oligidic diet weekly once

Table 3.5 Sex ratio and adult longevity (in days) of *Rhynocoris marginatus* reared on artificial diet with *Corcyra cephalonica* (CC) (T_1) and oligidic diets (AD) only (T_3)

Diets	Adult longevity		Sex ratio
	Male	Femle	
AD			
4	78.3 ± 4.5	83.0 ± 4.6*	0.76
5F1	85.7 ± 2.0	104.4 ± 6.5*	0.74
5F2	94.2 ± 4.7	108.3 ± 5.9*	0.67
6F1	85.0 ± 3.0	97.2 ± 4.1*	0.62
6F2	63.1 ± 2.2	74.1 ± 7.4*	0.79
6F3	76.5 ± 5.5	109.4 ± 2.3*	0.68
AD + C C			
4	77.2 ± 5.1	84.6 ± 9.1*	0.62
5F1	81.2 ± 4.9	84.5 ± 3.3*	0.72
5F2	78.0 ± 4.8	87.3 ± 5.9*	0.67
6F1	82.0 ± 2.9	88.9 ± 4.5*	0.76
6F2	75.2 ± 2.8	87.3 ± 6.4*	0.75
6F3	86.2 ± 3.2	104.5 ± 8.2*	0.76

F1, F2 and F3 stands for first, second, and third filial generations, respectively; '*t*' test $p < 0.05$ % level
*indicates significance at 5 % level

with *Corcyra cephalonica* sex ranged from 0.62 to 0.76. (chi square = 0.29, df = 4, $p = 0.99$) (Table 3.5). But in control category, the sex ratio ranged from 0.65 to 0.82 (chi square = 0.47, df = 5, $p = 0.99$) (Table 3.4).

Provision of water along with *Corcyra cephalonica* enhanced the female-biased sex ratio from 0.74 to 0.78 (chi square = 0.92, df = 4, $p = 0.92$). Similarly sex ratio was also enhanced in *Spodoptera litura* offered to *Rhynocoris marginatus* (0.78–0.85) (chi square = 0.85, df = 4, $p = 0.95$) (Table 3.4). The sex ratio is female biased in all the tested categories. It was ranged from 0.62 to 0.85. Similar female biased sex ratio was observed for *Rhynocoris marginatus* (Ambrose 1999; Claver et al. 1996). Moreover, Long and Zaher (1958) reported that insect reared at different diets produced maximum adults with greater effects on female. Field survey also revealed that population of female was higher than males.

The longevity of male and females was not significantly affected by diets with exception of the CW (T_2) male (55.8 days) (Table 3.4). In general females lived longer than the males. In control (*Corcyra cephalonica*) category females and males lived a maximum of 122 and 102 days, respectively. It has reduced when water was provided with *Corcyra cephalonica* in average of 55.8 days for male and 80.5 days for female. Similar observations were also observed when *Rhynocoris marginatus* was provided with *Spodoptera litura,* and OD + *Corcyra cephalonica* too (Table 3.5). The adult longevity was slightly longer when compared with natural prey *Spodoptera litura* (average of 80.13 days for male, and 92.63 days for female) or factitious host *Corcyra cephalonica* (average 106.6 days for male and 118.33 day-sfor female).

Table 3.6 Preoviposition (PRE), oviposition (OVI), and post oviposition periods (POVI) (in day), number of eggs laid (NEL), number of egg batches (NEB) and number of nymphs hatched (NNH) by *Rhynocoris marginatus* reared on *Corcyra cephalonica* and *Corcyra cephalonica* with water

Diets	PRE	OVI	POVI	NEL	NEB	NNH
Spodoptera litura						
SLF1	20.8 ± 2.1	28.8 ± 2.1*	15.8 ± 3.3*	131.8 ± 7.5*	3.2 ± 0.1*	94.5*
SLF2	19.8 ± 3.3	32.5 ± 3.5*	14.7 ± 2.2*	122.6 ± 6.8*	3.1 ± 0.2*	95.2*
SLF3	25.3 ± 2.1	39.3 ± 4.2*	12.3 ± 3.4*	108.2 ± 8.2*	3.1 ± 0.5*	91.2*
Corcyra cephalonica						
CCF1	26.2 ± 0.4	50.3 ± 0.8*	24.1 ± 0.3*	177.0 ± 8.6*	3.2 ± 0.1*	85.7*
CCF2	20.5 ± 1.1	95.0 ± 1.5*	29.5 ± 0.9*	88.7 ± 6.5*	2.9 ± 0.1*	88.9*
CCF3	22.1 ± 1.1	94.1 ± 2.2*	32.2 ± 3.2*	92.2 ± 1.2*	2.6 ± 0.2*	96.2*
Corcyra cephalonica + water						
CWF1	30.0 ± 1.0	56.9 ± 1.9*	26.9 ± 0.8*	170.9 ± 12.0*	2.7 ± 0.1*	87.0*
CWF2	22.2 ± 0.9	110.0 ± 1.1*	27.3 ± 0.8*	95.5 ± 4.8*	2.7 ± 0.1*	94.7*
CWF3	23.6 ± 1.4	87.0 ± 1.2*	23.0 ± 1.4*	105.4 ± 5.5*	2.1 ± 0.2*	96.4*

F1, F2 and F3 stands for first, second, and third filial generations, respectively; '*t*' test $p < 0.05$ % level
*indicates significance at 5 % level

3.3.6 Reproduction

Adults were also maintained on the same diet as in their nymphal instars. For each diet, 25 pairs were collected from cultures and maintained in 500 ml capacity plastic box (5.5 cm length and 12.5 cm diameter). Filter paper was furnished on the bottom of the container. Oligidic diet and, natural and factitious hosts were provided and excess and unconsumed preys were replaced every day. Preoviposition, oviposition and post-oviposition period and, total number of eggs laid per female and egg hatched on each category were monitored on a daily basis. The experiments were carried out in environmental chambers for three generations (Remi, Mumbai) continuously at 28 ± 1 °C, a relative humidity of 75 ± 80 % and a photoperiod of 11:13 (L:D) h.

Female adult longevity was divided into three distinct phases namely preoviposition, oviposition, and post-oviposition periods. In *Corcyra cephalonica* category, preoviposition period ranged from 20.5 to 26.2.0 days (average of 22.93 days). It was slightly prolonged (1–3 days) when water was provided along with the *Corcyra cephalonica* from 22.2 to 30 days ($t = 16.43$, df-13, $P = 0.0005$) (average of 25.26 days) (Table 3.6). In contrast, preoviposition period was similar for female fed with *Spodoptera litura* and *Corcyra cephalonica.* However, the oviposition period of *Rhynocoris marginatus* was lowered when it was fed with *Spodoptera litura* (average of 79.8 egg/female for CC T1, 84.63 for CW T2 and 33.53 egg/female for SL T5) (Table 3.6). It was also revealed from results that oviposition period was more or less similar in T3-and T4-reared *Rhynocoris marginatus* than the control category (Table 3.6). This was further changed when *Corcyra cephalonica* was provided along with the OD (78.56 days).

Table 3.7 Preoviposition (PRE), oviposition (OVI) and post oviposition periods (POVI) (in days), number of eggs laid (NEL), number of egg batches (NEB) and number of nymphs hatched (NNH) of *Rhynocoris marginatus* reared on different oligidic diets (T_3 and T_4)

Diets	PRE	OVI	POVI	NEL	NEB	NNH
Artificial diets						
4	28.1 ± 1.5	–	–	10.5 ± 0.7	1.0 ± 0.0	–
5F1	29.5 ± 1.1	76.0 ± 0.9*	26.3 ± 2.0*	19.0 ± 1.5*	1.5 ± 0.1	89.3*
5F2	25.2 ± 1.5	87.0 ± 0.1*	34.8 ± 1.5*	34.6 ± 3.5*	1.7 ± 0.1*	94.0*
6F1	26.5 ± 1.0	76.3 ± 2.1*	22.8 ± 1.2*	38.9 ± 4.9*	1.5 ± 0.1	92.3*
6F2	27.2 ± 2.2	65.3 ± 4.6*	22.1 ± 2.2*	46.0 ± 7.2	1.3 ± 1.2	76.7*
6F3	47.0 ± 7.3	58.2 ± 2.2*	26.1 ± 4.2*	53.0 ± 1.0*	1.3 ± 0.3*	95.8*
Oligidic diets + *C. cephalonica*						
4	27.0 ± 2.1	–	–	17.6 ± 2.6	1.0 ± 0.0	–
5F1	29.0 ± 1.3	82.1 ± 2.2*	16.4 ± 1.2*	30.2 ± 2.2*	1.5 ± 0.1*	83.8
5F2	25.7 ± 1.4	132.0 ± 2.5*	36.7 ± 3.0*	41.2 ± 3.1*	2.0 ± 0.1	94.5*
6F1	26.5 ± 1.2	86.0 ± 2.3*	16.4 ± 1.2*	47.7 ± 2.9	1.5 ± 0.1*	93.8
6F2	24.2 ± 3.1	54.4 ± 3.2*	23.0 ± 1.4*	61.0 ± 4.5	1.2 ± 0.2	74.6*
6F3	31.2 ± 2.3	38.3 ± 4.2*	13.0 ± 3.2*	59.6 ± 4.7*	1.7 ± 0.2*	94.1*

F1, F2 and F3 stands for first, second, and third filial generations, respectively '*t*' test $p < 0.05$ % level
*indicates significance at 5 % level

The results showed that the mean number of eggs laid by female *Crysoperla carnea* fed with diets containing egg yolk, vitamin "C" and mixed yeast extracts was 33.66% for AD (T_4). (Average for three generations) ($t = 21.62$, df = 12, $P = 0.0005$) and 42.88% for OC (T_3) (average for three) ($t = 14.54$, df = 12, $P = 0.0005$) and the CW (T_2) ($t = 11.40$, df = 12, $P = 0.0005$).

Rhynocoris marginatus laid 177.0 eggs when *Corcyra cephalonica* was provided as prey. It was reduced (170.9) egg/female when water was provided along with *Corcyra cephalonica.* Further decrease was observed when *Spodoptera litura* was provided (131.8 egg/female) (Table 3.6). When, *Rhynocoris marginatus* was reared with OD, initially the fecundity was very meager (5–10 eggs/female) then it was increased to diet 5 (19.0 and 34.6 egg/female for F1 and F2 generation, respectively). Further increase was recorded in diet- 6 (53.00 egg/female) (Table 3.7).

The mean number of eggs laid was higher in control categories, whereas fecundity was not significantly different for the prey *Spodoptera litura* (T5), *Corcyra cephalonica* (T1), *Corcyra cephalonica* weekly once with water (T2), respectively.

Sujatha and Sahayaraj (2007) also checked the influence of oligidic diet and factitious host on the development, survival and adult longevity of *Rhynocoris marginatus*. The incubation period of *Rhynocoris marginatus* egg was 8 days in the control category. The hatching percentage was gradually decreased with increasing the days of incubation. The percentage of hatched eggs was higher (average of 93.6 for T_5 and 90.26 for T_1) in the predator reared on natural prey than in the meat-based diet (89.56 and 88.16 for OD and OC, respectively). In AD and OD + CC categories, *Rhynocoris marginatus* laid eggs but they were not hatched. Generally the reproductive rate and hatchability of the predators reared on OD were less than those reared on factitious or natural prey *Spodoptera litura* (Tables 3.6 and 3.7). When *Corcyra cephalonica* and *Spodoptera litura* were provided, *Rhynocoris marginatus* females mate within 27 days of adult life and preoviposition period was slightly (2–3 days) affected by oligidic diet taking 28–30 days.

Meat-reared females of *Dicyphus tamanini* Waganer (Miridae) did not increase their preovipositional time. In contrast, De Clercq and Deghlee (1992) found an increase in preoviposition period when *Podisus* spp. were continuously reared on the meat diet as observed in this study. Also oviposition period was shorter (30–44 days) when *Rhynocoris marginatus* was reared on *Spodoptera litura*. It was prolonged from 51 to 97 days on *Corcyra cephalonica*. Further prolongation was recorded in diet-5 (up to 135 days), but reduced in diet 6. The rate of oviposition of *Rhynocoris marginatus* species can vary with the species of prey (Sahayaraj et al. 2004). In an average both *Spodoptera litura* and *Corcyra cephalonica* provided *Rhynocoris marginatus* laid 120 eggs/female.

It was 2.65 time reduced (56.1 egg/female) when the reduviid was reared with meat-based oligidic diet. Generally the reproductive rate of predators reared on artificial diets was high (Carpenter and Greany 1998; Cohen 1985a, b, 1992, 2000; Cohen and Staten 1994; Coudron et al. 2002; De Clercq et al. 1998; Wittmeyer and Coudron 2001; Rojas et al. 2000). Cohen and Smith (1998) reported significant cost savings in producing high-quality populations of the predator *Chrysoperla filabris*. When the predator *Panthous bioculatus* was maintained an artificial diet and reduced egg production were observed and were attributed to the failure to form mature follicles (Adams 2000). Our study demonstrated that addition of vitamin E, blood serum, and egg yolk enhanced the reproduction of *Rhynocoris marginatus*. Vitamin A, niacin, riboflavin B12, pantothenic acid, thiamin, pyridoxine, folic acid, and vitamins E and D are present in greater quantity in egg yolk.

Similarly folic acid, which is particularly more important for egg productions is much higher (117 g) in egg yolk. Egg yolk also has higher amount of saturated, mono unsaturated, polyunsaturated oils and lipids. Furthermore, the egg yolk has greater calorific value (303 cal per 100 g). The cholesterol level is particularly very high (1075 mg) in egg yolk (says Rolfes et al. 1978). Diet containing egg yolk is quite rich in proteins, minerals, vitamins and lipids as compared to the diets containing egg white and mixed egg (says Rolfes et al. 1978; Norioka et al. 1984), which promoted quick growth and completion of the larval period. It was also reported that after pre-oviposition, feeding during adult phase plays a more important role than feeding during the nymphal phase in terms of egg production (Cangussu and Zucoloto 1992, 1995; Fernandes-da-Silva and Zucoloto 1993). The result obtained did not support the above view because it has not either provided any special food and/or prey or taken any special care during the preoviposition period.

Lower fecundity observed in predators reared on artificial diet may there fore be partially attributed to lower adult weights. Female with lower body weights did not however adjust the weight of individual eggs. They decreased reproductive output rather by reducing oviposition frequently total number of egg deposited and size of egg batches. These findings are consistent with those of O'Neil and Wiedenmann (1990) who investigated the effects of feeding regimens on reproduction.

A quantitative examination of eggs laid though the oviposition period was done by Ferkovich and Shapiro (2004a, b) who suggested that the increase in egg deposition on multiple mated females may be related to hormonal effects in egg production. Though multiple mating was possible also recorded in this study, the total number of eggs laid and also the number of egg batches laid by *Rhynocoris marginatus* were so poor. This might be due to the lower nutritional value, alternation of gut bacterial population and their enzyme and protein profiles recorded in *Rhynocoris marginatus*. It has been observed that repeatedly mated females had short lives (Lamunyon 1997).

It also indicates that some nutrient may inhibit the conversion of *Corcyra cephalonica* in to body mass and also prevents the oogenesis of *Rhynocoris marginatus*. We hypotheized that if we alter the ingredients of OD and also rearing medium of *Corcyra cephalonica* the reproductive ability can be increased. Previously (Sahayaraj and Sathyamoorthi 2002) recorded that change of *Corcyra cephalonica* rearing media, could alter the reproductive ability of *Rhynocoris marginatus*. If the chemical composition of OD was changed, dependently the insect behavior, biology, reproduction, and morphology. Cohen (1992) pointed out that besides the evaluation of behavior performance, quality control of predators reared on artificial diet should also include immunological procedures and metabolic tests. Such tests may result in a better understanding of the trophic biology of the predators, and may consequently allow for making appropriate dietary adjustments to improve the nutritional value of the artificial diet.

Development of *Rhynocoris marginatus* in the meat—based was slower and female weight, survival, fecundities, and oviposition periods were also lower. However, this indicates that the insects were smaller and they were able to reproduce as successfully as conventionally reared insects when offered this same conditions. Normal values of female longevity and oviposition rate as well as the total number of eggs produced by females with these diets show the potential of *in vitro* rearing of this predator. Also, the failure of some individuals reared on artificial diets to produce normal adult did not affect the performance of the first generation adults (De Clercq and Degheele 1997).

Lower reproduction in AD category might also be due to lower rates of assimilation and conversion efficiency as observed by Cohen (1989, 1992), and Ferkovich and Shapiro (2003). It may also be due to the failure of follicle maturity (Adams 2000) or reduced juvenile hormone titer or lack of vitellogenin precursors such as amino acids, lipids, or carbohydrates (Ferkovich and Shapiro 2004). The author attributed the reduction in egg production not only to a protein deficiency but also to a lack of chemical or behavioral cues that confirmed the presence of live prey.

A nutritional deficiency in the artificial diet is likely to suppress *Rhynocoris marginatus* in succeeding generations. This suggests that larger nymphs and adults of *Rhynocoris marginatus* have difficulties handling this type of inanimate food, probably because of the low amount of nutrients, and can not extract nutrient at an appropriate rate to fully support growth and reproduction. We recorded abnormalities of both nymph and adults of *Rhynocoris marginatus* while they are reared on diet 1 to diet 4. Similarly, abnormal may indicate a nutritional problem (Craig

1997). This result closely corresponds to the presentation for *Macrolophus caliginosus* reared on a similar meat-based diet (Castane and Zapata 2005; Iriarte and Castane 2001).

Females of many insect species may obtain extra nutrients and water through mating (Marshall and McNeil 1989; Pivnick and McNeil 1987). During mating, the male transfers spermatophore which contains sperm, and male accessory gland fluid of male insects in general but, for some lepidopterns (Drummond 1984; Greenfield 1982) they were transferred during copulations. Our results showed that provision of water along with either *Corcyra cephalonica* or artificial diets shows no direct effects on reduviid ovipositions.

This study clearly shows that although *Rhynocoris marginatus* can be reared on an oligidic diet that is economically produce, the fecundity of the female is significantly less than those reared on either *Corcyra cephalonica* or *Spodoptera litura.* In the current study, the oligidic diet, originally developed for these predatory reduviids, proved suitable food for sustainable development and reproduction of *Rhynocoris marginatus* reared either individually or in group. Further experiments will investigate whether this diet will sustain <10 generations of *Rhynocoris marginatus*, given that nutritional deficiencies may be expressed only after several generations (Arijs and De Clercq 2004).

3.3.7 Life Table Parameters

By using oviposition data, life tables were constructed according to the methods recommended by Birch (1948) and modified by Southwood (1978). In life table statistics, the intrinsic rate of increase has been determined using the equation '$\sum e^{-rm \times l_x m_x - 1}$', where *e* is the base of natural logarithms, *x* is the age of the individuals in days, '*lx*' is the number of individuals alive at age *x* as the proportion of *1* and *mx* is the number of female offspring produced per female in the age interval '*x*'. The sum of products '*lxmx*' is the net reproductive rate 'R_0'. The rate of multiplication of population for each generation has been measured in terms of females produced per generation. The precise value of cohort generation has been calculated as follows:

$$\text{Mean length of generation } T_c = \frac{\sum l_x m_x}{R_0}$$

The arbitrary value of innate capacity for increase "*rc*" has been calculated applying the equation innate capacity for increase for increase in number

$$r_c = \frac{\log_e R_0}{T_c}$$

This is an appropriate rm value. The values of negative exponent of "*e −rmx*" ascertained from this experiment often lay outside the range. For this reason both sides of the equation have been multiplied by a factor of "$\sum e^{7-rm \times l_x m_x}$"—1096.6 (Birch 1948; Watson 1964). The two values of "$\sum e^{7-rm \times l_x m_x}$" have been then plotted on the horizontal axis against their respective arbitrary '*rm*' on the vertical axis.

Two points have been then joined to give a line, which is intersected by a vertical line drawn from the desired value of "$e^{-rm \times l_x m_x}$" (1096.6).

The point of intersection gives the value of '*rm*' accurate to three decimal places. The precise generation time '*T*' has been then calculated from the equation

$$T = \frac{\log_e R_0}{R_m}$$

The finite rate of increase (λ) has been calculated as "*erm*". This "λ" represents the number of individuals added to the population per female per day (Siddiqui et al. 1973). The weekly multiplication of predator population has been calculated as "erm^7". The doubling time has been calculated as log '2/log λ'.

Analyses of life table showed that the net reproductive rate (NRR), arbitrary value of innate capacity for increase, innate rate of increase and weekly multiplication were maximum in *Corcyra cephalonica* and *Corcyra cephalonica* + water (Table 3.8). These parameters were gradually decreased in oligidic diets-6 and 5 and also oligidic diet and weekly once with *Corcyra cephalonica* (5C and 6C) (Table 3.8). Precise generation time was minimum (14.58 days for diet 5A) and maximum for oligidic diet (6A F3), respectively.

The life table parameters of *Rhynocoris marginatus* on five categories showed that (Tables 3.8 and 3.9) the net reproductive rate was significantly higher in *Spodoptera litura* (81.50, average 74.83) reared predators than other diets like *Corcyra cephalonica* (78.00) and OD (22.00 for diet 5) categories. The corrected rm population was higher for *Corcyra cephalonica* (average 0.52) followed by *Spodoptera litura* (0.48) and 4A to 6A (1.00, 0.96, 1.51, and 0.90, respectively). The precise generation time (T), finite rate of increase, and weekly multiplication time were always greater in *Spodoptera litura* than other diets.

However, more or less similar results were observed in *Corcyra cephalonica.* The doubling time (C) ranged from 1.00 to 1.47 for F1 to F3 *Corcyra cephalonica.* The hypothetical F2 female was maximum in T_2 category (4624.00) and minimum in 5A (484.00) category. Variation in the quantity of nutrients of prey species appears to have considerable effect on the feeding efficiency and reproductive potential of the predators (Beddington 1975). It reveals that *Spodoptera litura* and *Corcyra cephalonica* might be due to the minimum stress developed during feeding on less number of preys due to the comparatively larger size with richer body tissue. Awadallah et al. (1986) and George et al. (1998) reported same results in reduviids with five different preys. In all diets (From T_1 to T_5) life table statistics varied with the diet. Natality and survivorship curves reflected the higher on reproductive rate (R_0) of *Rhynocoris marginatus* on *Spodoptera litura* than on other categories. The mean length of generation was shorter for OD (T_3)-reared *Rhynocoris marginatus* followed by T_4 and T_3 (CW). The decrease was corrected rm and extension of developmental period, the population of doubling time increased in *Spodoptera litura* categories followed by T_1 and T_3 categories. Similarly finite rate of increase and doubling time were observed for *Rhynocoris marginatus* of different prey species (George 2000; Ravi 2004) and for *Sphedanolestes minusculus* (Bergroth).

Table 3.8 Life table parameters of *Rhynocoris marginatus* on different diets 4–6 and *Corcyra cephalonica* with Oligidic diet

Parameters	Oligidic diet						*Corcyra cephalonica* with oligidic diet					
	4	5F1	5F2	6F1	6F2	6F3	4C	5F1	5F2	6F1	6F2	6F3
Net reproductive rate (NRR)	–	22	23	33.25	48.70	52.25	28	44	60	62.0	64.75	65.75
Mean length of generation (Tc)	32.5	14.63	25.82	46.75	63.25	64.46	30.82	45.72	62	56.06	67.41	88.57
Innate capacity for natural increase (rc)	0.92	1.50	0.89	0.052	0.055	0.062	0.908	0.96	0.96	0.041	0.041	0.044
Corrected rm	0.96	1.51	0.90	0.055	0.045	0.065	0.910	0.97	0.97	0.042	0.042	0.038
Precise generation time (T)	31.25	14.58	25.55	56.75	57.22	58.71	30.76	45.36	61.85	77.28	85.96	86.25
Finite rate of increase (λ)	1.84	3.02	1.79	1.04	1.05	1.07	1.81	1.92	1.92	1.049	1.050	1.046
Weekly multiplication time	1.92	3.76	3.6	12.35	10.02	11.27	3.27	1.94	1.94	4.54	4.26	4.53
Doubling time (C)	1.08	0.66	1.11	1.45	1.52	1.60	1.02	1.04	3.84	1.395	1.329	1.389
Hypothetical female in F2 generation	900	484	529	862.75	982.52	1943.06	784	1936	3600	4356.0	4526.25	4423.06

Table 3.9 Life table parameters of *Rhynocoris marginatus* on *Corcyra cephalonica* (T_1) *Spodoptera litura* (T_5) and *Corcyra cephalonica* with water (T_3)

Parameters	*Corcyra cephalonica*			*Spodoptera litura*			*Corcyra cephalonica* with water		
	CC F1	CC F2	CC F3	SLF1	SLF2	SLF3	CW F1	CW F2	CW F3
Net reproductive rate (NRR)	78.00	73.5	66.5	81.50	72.75	70.25	38.00	31.00	68.00
Mean length of generation (Tc)	78.55	95.18	95.44	89.40	91.87	85.6	39.21	32.35	68.52
Innate capacity for natural increase (rc)	0.99	0.44	0.13	0.050	0.045	0.051	0.96	0.96	0.99
Corrected rm	1.00	0.55	0.69	0.56	0.072	0.055	0.98	0.97	1.00
Precise generation time (T)	78.0	88.48	77.95	78.25	81.48	78.21	38.77	31.95	68.0
Finite rate of increase (λ)	2.00	1.51	1.50	1.57	1.061	1.056	1.92	1.92	1.98
Weekly multiplication time	1.98	13.55	14.62	11.41	12.72	11.95	1.96	1.94	2.00
Doubling time (C)	1.00	1.419	1.409	1.477	1.258	1.569	1.04	1.04	3.96
Hypothetical female in F2	1260.84	53,658.2	4552.23	6881.56	5025.54	4954.22	1444	961	4624

Among the five diets, *Spodoptera litura* and *Corcyra cephalonica* for the following reasons showed faster development with higher survival, higher fecundity, higher net reproductive rate, and doubling time of the higher predator *Rhynocoris marginatus*. The above mentioned parameters were low in OD-, OC-, and CW reared predators. High net reproductive rate and intrinsic rate of population increase have been reported for temperate bugs such as *Oncopeltus guildnii,* and *Clavigrlla tomentosicollis* (Iheagwam 1982) when reared on their preferred hosts. The net reproduction and intrnsic rate of increase were greater for conventionally reared than for oligidic diet-reared *Macrolophus caliginosus* (Castane and Zapata 2005) as observed in this study.

3.4 Demerits of the Artificial Diets

For some time, the predators entangle with the feeding arena i.e., cotton, and are unable to come out because of the sticky nature. During the course of time, fungi grow, cause death and transformed in to cadavers. In another occasion, during feeding animals have the chance to move across the diet, which is to be the source of medium for microbe's growth and reproduction. During the course of time, microbes kill the insects. In support, Zhang et al. (2008) reported that although most artificial diets have been made in liquid form due to ease of mixture and insect ingestion, the adhesive nature of the material also increases mortality (Plate 3.1).

Sahayaraj and Sujatha's (2012) results showed the deformities in the cephalic appendages of forelegs, midlegs and hind legs. These deformities were more pronounced in fifth stadium than other nymphal stadia. When we reared *Rhynocoris marginatus* with diets 3, 4, and 5, both nymphs (Plate 3.2c, d) and adults (Plate 3.2e–f) were having deformities. Their kinds of deformities were common in *Rhynocoris marginatus*. They were: malformed legs (Plate 3.2c–f), curved and/or folded wings and exuvia present along with adults. If the AD soaked was not removed, it induces disease infection on *Rhynocoris marginatus* nymphs (Plate 3.2a) and adults (Plate 3.2b).

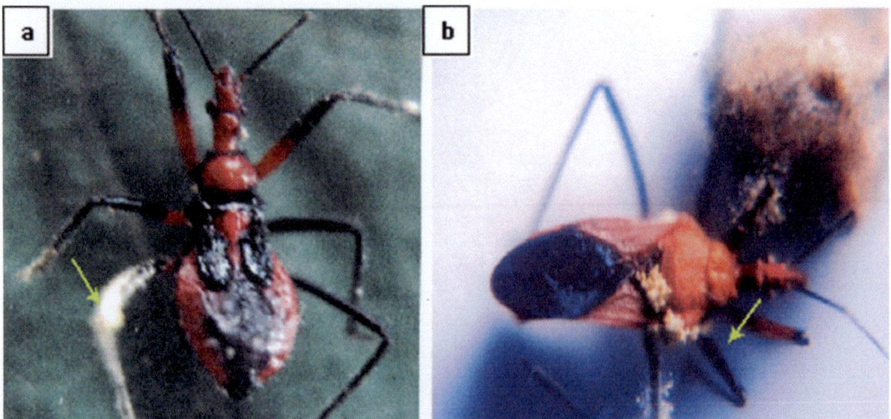

Plate 3.1 *Rhynocoris marginats* nymph (**a**) and adult (**b**) entangled in artificial diet which leads to microbial infection which latter cause death of the animal

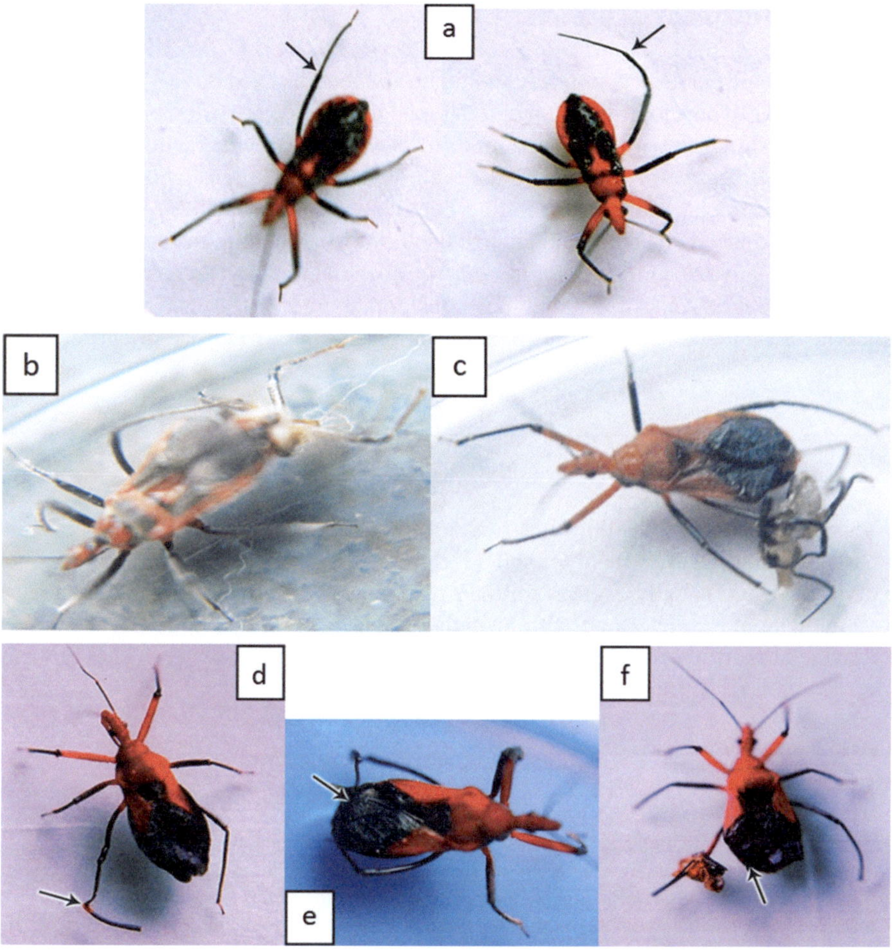

Plate 3.2 Deformities in *Rhynocoris marginatus* young ones (**a**) and adults (**c–f**) and adult of *Rhynocoris fuscipes* reared with artificial diet. Deformed legs (**a**, **d**, **e**, **f**), molted skin attached to the last segment of the abdomen (**b**) or legs (**c**, **f**), and deformed wings (**e**)

All of the diets had nutritional qualities that allowed complete development of the predators to a greater or a lesser extent, indicating that there is a potential for rearing this insect on artificial media and that the OD diet used adequate, allowing the nymph to acquire food and develop into the adult stage (De Clercq et al. 1998). Invariably the sex ratio was female biased for all reduviids. Long and Zaher (1958) reported that insect reared on different diets produced maximum adults with greater effects on female. Field survey also revealed that invariably the population of female was higher than of males. It had showed many problems that aroused during out experiments. Further batch to batch defects are due to:

1. The undefined nutrients and in adequate bioavailability
2. The diet does not truly replace the target insect natural food

3.5 Conclusions

Meat-based artificial diets not only elucidate the feeding of reduviid predators but also support the development of the predators. Our study demonstrates that reduviid predator feeding behavior cannot be modified if they are provided with artificial diet as food. Differences in the nymphal development and survival demonstrate that these predators can adapt their feeding behavior according to the food provided. Since the proposed diet could not change the total nymphal developmental period of *Rhynocoris kumarii* and *Spodoptera collaris*, artificial rearing can be carried out on the proposed diet. An artificial diet may considerably reduce the costs of mass propagation of beneficial insects and these meat-based diets were cheap and easy to prepare.

3.6 Future Focus

- Modulate the composition of artificial diet to enhance the reproduction, adult longevity and reduce the nymphal developmental time and mortality
- More reduviids will be tested for their response against the artificial diet
- Supplementary diet along with artificial diet will be provided and evaluate the artificial rearing of reduviid could be studied.

References

Adams TS (2000) Effect of diet and mating status on ovarian development in a predaceous stink bug *Perillus bioculatus* (F.) (Heteroptera: Pentatomidae). Ann Entomol Soc Am 93:529–535

Ambrose DP (1999) Assassin bugs. Science Publishers, Enfield

Ambrose DP, Saju T, Sahayaraj K (1990) Prey influence on the development, reproduction and size of the assassin bug *Rhynocoris marginatus*. Mitteilungen aus dem Museum fur Naturkunde in Berlin. Zoologisches Museum und Institut fur Spezielle Zoologie (Berlin), 8(1):280–287

Ambrose DP, Kulandaisamy AMJ, Sahayaraj K (1994) Impact of starvation, eye blinding and antennectomy on the predatory behaviour and efficiency of *Euagoras plagiatus* Burmeister. Bull Entomol 35(1–2):90–100

Arijs Y, De Clercq P (2001) Rearing *Orius laevigatus* on cysts of the brine shrimp *Artemia franciscana*. Biol Control 21:79–83

Arijs Y, De Clercq P (2002) Artificial diets for the production of natural enemies (predators and parasitoids) of greenhouse pest insects. Final consolidated report, FAIR-project nr. CT-98 4322

Arijs Y, De Clercq P (2004) Liver-based artificial diets for the production of *Orius laevigatus*. BioControl 49(5):505–516

Awadallah KT, Tawfik MFS, Husseini MM, Afifi AI (1986) Biocycle of the anthocorid predator *Xylocoris flavipes* in association with rearing on major pests of stored drug materials. Bull Soc Entomol Egypt 66:27–33

Beddington JR (1975) Mutual interference between parasites or predators and its effect on searching efficiency. J Anim Ecol 44:331–340

Birch S (1948) The intrinsic rate of natural increase in an insect populations. J Anim Ecol 17:15–26

Bradshaw WE, Johnson K (1995) Initiation of metamorphosis in the pitcher-plant mosquito: effects of larval growth history. Ecology 76:2055–2065

Cangussu JA, Zucoloto FS (1992) Nutritional value and selection of different diets by adult *Ceratitis capitata* flies (Diptera: Tephritidae). J Insect Physiol 38:485–491

Cangussu JA, Zucoloto FS (1995) Self-selection and perception threshold in adult females of *Ceratitis capitata* flies (Diptera : Tephritidae). J Insect Physiol 41:223–227

Carpenter JE, Greany PD (1998) Comparative development and performance of artificially reared versus host-reared *Diapetimorpha introita* (Cresson) (Hymenoptera: Ichneumonidae) wasps. Biol Control 12:137–142

Castane C, Zalom FG (1994) Artificial oviposition substrate for rearing *Orius insidiosus* (Hemiptera, Anthocoridae). Biol Control 4(1):88–91

Castane C, Zapata R (2005) Rearing the predatory bug *Macrolophus caliginosus* on a meat-based diet. Biol Control 34:66–72

Claver MA, Rajan K, Ambrose DP (1996) Impact of mass rearing in the post embryonic development of *Rhynocoris kumarii* Ambrose and Livingston (Heteroptera: Reduviidae). In: Ambrose DP (ed) Biological and cultural control of insect pests, an Indian cenario. Adeliine Publishers, Tirunelveli, pp 216–219

Cohen AC (1985a) Simple method for rearing the insect predator *Geocoris punctipes* (Heteroptera: Lygaeidae) on a meat diet. J Econ Entomol 78:1173–1175

Cohen AC (1985b) Metabolic rates of two hemipteran members of a predator-prey complex. Comp Biochem Physiol 81:833–836

Cohen AC (1989) Ingestion efficiency and protein consumption by a heteropteran predator. Ann Entomol Soc Am 82:495–499

Cohen AC (1990) Feeding adaptations of some predaceous Hemiptera. Ann Entomol Soc Am 83(6):1215–1223

Cohen AC (1992) Using a systematic approach to develop artificial diets for predators. In: Anderson TE, Leppla NC (eds) Advances in insect rearing for research and pest management. Westview Press, Boulder, pp 77–91

Cohen AC (1995) Extra-oral digestion in predaceous terrestrial Arthropoda. Annu Rev Entomol 40:85–103

Cohen AC (2000) Feeding fitness and quality of domesticated and feral predators: effects of long-term rearing on artificial diet. Biol Control 13:49–54

Cohen AC, Smith LK (1998) A new concept in artificial diets for *Chrysoperla rufilabris*: the efficacy of solid diets. Biol Control 13:49–54

Cohen AC, Staten RT (1994) Long term culturing and quality assessment of redatory big-eyed bugs, *Geocoris punctipes*. In: Narang SK, Bartlett AC, Faust R (eds) Applied genetics of biological control. CRC Press, Boca Raton, pp 121–132

Cohen AC, Urias NM (1986a) Meat-based artificial diets for *Geocoris punctipes*. Southwest Entomol 11:171–176

Cohen AC, Urias NM (1986b) Meat-based artificial diets for *Geocoris punctipes* (Say). Southwest Entomol 11:171–176

Coudron TA, Kim Y (2004) Life history and cost analysis for continuous rearing of *Perillus bioculatus* (Heteroptera: Pentatomidae) on a zoophytogenous artificial diet. J Econ Entomol 97:807–812

Coudron TA, Wittmeyer J, Kin Y (2002) Life history and cost analysis for continuous rearing of *Podisus maculiventris* (Say) (Heteroptera : Pentatomidae) on a zoophytophagous artificial diet. J Econ Entomol 95:1159–1168

Craig CL (1997) Evolution of arthropod silks. Annu Rev Entomol 42:231–267

De Clercq P, Degheele D (1992) A meat-based diet for rearing the predatory stinkbugs *Podisus maculiventris* and *Podisus sagitta* (Hetroptera : Pentatomidae). Entomophaga 37(1):149–157

De Clercq P, Degheele D (1993) Quality assessment of the predatory bugs *Podisus maculiventris* (Say) and *Podisus sagitta* (Fab.) (Heteroptera: Pentatomidae) after prolonged rearing on a meat—based artificial diet. Biocontrol Sci Tech 3(2):133–139

De Clercq P, Degheele D (1997) Effects of mating status on body weight, oviposition, egg load, and predation in the predatory stinkbug *Podisus maculiventris* (Heteroptera: Pentatomidae). Annu Entomol Soc Am 90:121–127

De Clercq P, Merlevede F, Tirry L (1998) Unnatural prey and artificial diets for rearing *Podisus maculiventris* (Heteroptera: Pentatomidae). Biol Control 12(2):137–142

Drummond BA (1984) Multiple mating and sperm competition in the Lepidoptera. In: Smith RL (ed) Sperm competition. Academic Press, London, pp 291–370

Ferkovich S, Shapiro J (2003, September) Strategies for improving artificial diets with insect-derived components. In: Arthropod mass rearing and quality control workshop, pp 26

Ferkovich SM, Shapiro JP (2004a) Increased egg-laying in *Orius insidiosus* (Hemiptera: Anthocoridae) fed artificial diet supplemented with an embryonic cell line. Biol Control 31(1):11–15

Ferkovich SM, Shapiro JP (2004b) Comparison of prey-derived and non-insect supplements on egg-laying of *Orius insidiosus* maintained on artificial diet as adults. Biol Control 31:57–64

Fernandes-da-Silva PG, Zucoloto FS (1993) The influence of host nutritive value on the performance and food selection in of *Ceratitis capitata* (Diptera Tephritidae). J Insect Physiol 39:883–887

Firlej A, Chouinard G, Coderre D (2006) A meridic diet for the rearing of *Hyaliodes vitripennis* (Hemiptera: Miridae), a predator of mites in apple orchards. Biocontrol Sci Tech 16(7):743–751

Freitas SPC, Júnior E, Santos W, Zanuncio JC, Serrão JE (2006) Development, survival and reproduction of *Podisus nigrispinus* (Dallas, 1851)(Heteroptera: Pentatomidae) with salt and amino acids solutions supplementary diet. Braz Arch Biol Technol 49(3):449–455

George PJE (2000) Nymphal cannibalistic bahaviour of *Rhynocoris kumarii* ambrose and livingstone (Heteroptera: Reduviidae) to prey deprivation and mass rearing conditions. J Biol Control 14(1):1–3

George PJE, Seenivasagan R, Kannan S (1998) Influence of prey species on the development and reproduction of *Acanthaspis siva* Distant (Heteroptera: Reduviidae). Entomon 23(4):313–319

Greenfield MD (1982) The question of paternal investment in Lepidoptera : male contributed proteins in *Plodia interpunctella*. Int J Invertebr Reprod 5:323–330

Grenier S, De Clercq P (2003) Comparison of artificially versus naturally reared natural enemies and their potential for use in biological control. In: van Lenteren J (ed) Quality control and production of biological control agents: theory and testing procedures. CABI Publishing, Wallingford, pp 115–131

House HL (1967) Artificial diets for insects. A compilation of references and abstracts, Information Bulletin No. 5. Research Institute Canada Dept. Agric, Ontario

Iheagwam EU (1982) Effects on host plants (legume) on life and population parameters of *Clavigrella tomemtosicollis* (Hemiptera : Coreidae). Rev Zool Africane 96:912–920

Iriarte J, Castane C (2001) Artificial rearing of *Dicyphus tamaninii* (Heteroptera : Miridae) on a meat-based diet. Biol Control 22:98–102

Lamunyon C (1997) Increased fecundity, as a function of multiple mating, in an arctiid moth, Utetheisa ornatrix. Ecol Entomol 22(1):69–73

Lepis J, Travis J (1994) Metamorpic responses to changing food levels in two species of hylid frogs. Ecology 75:1345–1356

Long DP, Zaher MA (1958) Effect of larval population density on the adult morphology of two species of Lepidoptera, *Plusi gamma* L. and *Pieris brassicae* L. Entomol Exp Appl 1:161–173

Marshall LD, McNeil JN (1989) Spermatophore mass as an estimate of male nutrient investment:a closer look in *Pseudaletia unipuncta* (Haworth) (Lepidoptera : Noctuidae). Funct Ecol 3:605–612

Norioka N, Okada T, Hamazume Y, Mega T, Ikenaka T (1984) Comparison of nutritive value of egg yolk and egg white and whole egg. J Biochem 97:19–28

Nylin S, Gotthard K (1998) Plasticity in life history traits. Annu Rev Entomol 43:63–83

O'Neil RJ, Wiedenmann RN (1990) Body weight of *Podisus maculiventris* (Say) under various feeding regimes. Can Entomol 122:285–294

Pivnick KA, McNeil JN (1987) Puddling in butterflies: sodium affects reproductive success in *Thymelicus lineola*. Physiol Entomol 12:461–472

Ravi (2004) Interaction of selected reduviids and botanicals in groundnut pest management. Ph. D. Thesis. Tirunelveli, Manonmaniam Sundaranar University. p 74–85

Rojas MG, Morales-Ramos JA, King EG (2000) Two meridic diets for *Perillus bioculatus* (Heteroptera: Pentatomidae), a predator of *Leptinotar sadecemlineata* (Coleoptera: Chrysomelidae). Biol Control 17:92–99

Rolfes T, Clement P, Winter AR (1978) The chemical composition of eggs. J Sci Food Agric 17:101

Sahayaraj K (2002) Small-scale laboratory rearing of reduviid predator *Rhinocoris marginatus* (Fab.) (Hemiptera : Reduviidae) on *Corcyra cephalonica* Stainton larvae by larval card method. J Cent Eur Agric 3(2):137–148

Sahayaraj K (2007) Isolation, identification and characterization of gut flora of three reduviid predators. Asian J Microbiol Biotechnol Environ Sci 9(4):1073–1075

Sahayaraj K (2008) Approaching behaviour of *Rhynocoris marginatus* (Fab.) (Heteroptera: Reduviidae) on three prey kairomones. Bull Insectol 61(2):233–237

Sahayaraj K (2012) Artificial rearing on the nymphal developmental time and survival of three reduviid predators of Western Ghats, Tamil Nadu. J Biopest 5(2):218–221

Sahayaraj K, Paulraj MG (2001) Rearing and life table of reduviid predator *Rhynocoris marginatus* Fab. (Heteroptera : Reduviidae) on *Spodopter a litura* Fab. (Lepidoptera : Noctuidae) larvae. J Appl Entomol 125(6):321–325

Sahayaraj K, Sathyamoorthi P (2002) Influence of different diets of *Corcyra cephalonica* Stainton on life history of *Rhynocoris marginatus* (Fab.) (Hemiptera: Reduviidae). J Cent Eur Agric 3(1):53–62

Sahayaraj K, Sujatha S (2012) Temperature-dependent biology and physiology of predatory reduviids. Nova Publication, New York, p 179

Sahayaraj K, Martin P, Karthikraja S (2003) Suitable sex ratio for the mass rearing of reduviid predator *Rhynocoris marginatus* (Fab.). J Appl Zool Res 14(1):34–37

Sahayaraj K, Thangarani S, Delma JCR (2004) Comparative prey suitability of *Helicoverpa armigera* and *Spodoptera litura* larvae for *Rhynocoris marginatus* (FAB.) (Heteroptera: Reduviidae). Belgium J Entomol 6:383–392

Sahayaraj K, Martin P, Selvaraj P, Raju G (2006) Feeding behaviour of Reduviid predators on meat and insect – based artificial diets. Belgean J Entomol 8:55–65

Sahayaraj K, Venkatesh P, Balasubramanian R (2007a) Feeding behaviour and biology of a reduviid predator *Rhynocoris marginatus* (Fabricius) (Heteroptera: Reduviidae) on Oligidic Diet. Hexapoda 14(1):24–30

Sahayaraj K, Kumara Sankaralinkam S, Balasubramanian R (2007b) Prey influence on the salivary gland and gut enzymes qualitative profile of *Rhynocoris marginatus* (Fab.) and *Catamiarus brevipennis* (Serville) (Heteroptera: Reduviidae). J Insect Sci 4(4):331–336

Siddiqui WH, Barriow CA, Randolph PA (1973) Effect of some constant and alternating temperatures on population growth of the pea aphid *Acyrthosiphon pisum* (Homoptera : Aphididae). Can Entomol 105:145–146

Sikorowski PP, Goodwin RH (1985) Contaminant control and disease recognition in laboratory cultures. In: Singh P, Moore RF (eds) Handbook of insect rearing, vol I. Elsevier, New York, pp 85–105

Singh P (1977) Artificial diets for insects, mites and spiders. Plenum, New York, pp 8–12

Slansky F Jr, Rodriguez JG (1987) Nutritional ecology of insects, mites, spiders, and related invertebrates. Wiley, New York

Southwood TRE (1978) Ecological methods: with particular reference to the study of insect population. Chapman and Hall, London

Sujatha S, Sahayaraj K (2007) Influence of oligidic diet and factitious host on the development, survival and adult longevity of *Rhynocoris marginatus* (Fabricious) (Heteroptera: Reduviidae: Harpactorinae). In: Narayann M, Sethuramalingam T, Sahayaraj K (eds) National Seminar on Technology and Management of Bioresearches, p 35–39

Sundararaju D (1984) Studies on cashew pests and their natural enemies in Goa. J Plant Crop 12(1):38–48

Tan XL, Zhao J, Wang S, Zhang F (2015) Optimization and evaluation of microencapsulated artificial diet for mass rearing the predatory ladybird *Propylea japonica* (Coleoptera: Coccinellidae). Insect Sci 22(1):111–120

Thompson SN, Hagen KS (1999) Nutrition of entomophagous insects and other arthropods. In: Bellows TS, Fisher TW (eds) Biological control: principles and applications of biological control. Academic, San Diego, pp 594–652

Waage JK, Carl KP, Mills NJ, Greathead DJ (1985) Rearing entomophagous insects. In: Singh P, Moore RF (eds) Handbook of insect rearing. Elsevier, New York, pp 45–66

Watson TF (1964) Influence of host plant condition on population increase of *Tetrancychus telerium* (L.) (Acaina : Tetranychidae). Hilgardia 35:273–322

Wittmeyer JL, Coudron TA (2001) Life table parameters, reproductive rate, intrinsic rate of increase, and estimated cost of rearing *Podisus maculiventris* (Heteroptera : Pentatomidae) on artificial diet. J Econ Entomol 94(6):1344–1352

Zanuncio JC, Saavedra JLD, Oliveira HN, De Gheele D, De Clercq P (1996) Development of the predatory stinkbug *Brontocoris tabidus* signoret heteroptera: pentatomidae on different proportions of an artificial diet and pupae of *Tenebrio molitor* L. Coleoptera: Tenebrionidae. Biocontrol Sci Tech 6(4):619–626

Zhang SC, Zhou XM, Pan Y, Lei CL (2008) Evaluation of an artificial liquid diet of *Orius similis* Zheng (Hemiptera: Anthocoridae). Acta Entomol Sin 51:997–1001

Gut Autochthonous Microbes and Their Enzyme Profile

4

Abstract

Insects dependent on various diets commonly carry symbiotic microorganisms that provide nutritional supplements for their host. Since antimicrobial agent has been reared in oligidic diet to prevent microbial contamination, it is essential to study the microbiological nature of symbiotic bacteria on artificial diet-reared predators. Studies of associations between artificial diet natural prey feeding and gut microorganisms of insect predators have sharpened appreciation for utilizing the diet in predator rearing gained importance. In this study, it tested that artificial diet has any influence on microbial community composition in *Rhynocoris marginatus* adults compared with natural prey source. The study suggested that the total heterotrophic bacterial population (THBP) is varied in the diet ingestion by the reduviid *Rhynocoris marginatus*. Among all the categories, the bacterial population was found to be higher in *Spodoptera litura* category followed by *Corcyra cephalonica*, *Corcyra cephalonica* with water, and oligidic diet with *Corcyra cephalonica*. THBP was drastically decreased in artificial diet-reared reduviids. It showed that microbe's population was drastically altered if the food nature changed. Natural food like *Spodoptera litura* keeps the gut microbes viable and enriches gut natural population. A total of 14 bacterial species were recorded in artificial diet-reared *Rhynocoris marginatus*, followed by the *Corcyra cephalonica* (9) and *Spodoptera litura* (9). *Micrococcus* sp., *Bacillus subtilis*, *Pseudomonas fluorescens*, *Streptococcus faecalis*, and *Aeromonas* sp. were also isolated from the predators. *Micrococcus varians* was a dominant (52.70 %) species diet.

In natural prey-reared reduviid, the number of bacterial cells remained almost constant during the experiment, while the number of bacteria than other artificial diets increased. In contrast, the number of bacterial species within artificial diet-reared insect decreased. In natural prey, *Spodoptera litura*, nine bacteria were observed with varying numbers of bacterial cells. In T1 categories, ten bacteria species were isolated followed by species in T3 categories (artificial diet). Fourteen bacterial species were isolated from T4 category (artificial diets + *Corcyra cephalonica*).

K. Sahayaraj, R. Balasubramanian, *Artificial Rearing of Reduviid Predators for Pest Management*, DOI 10.1007/978-981-10-2522-8_4

Keywords
Insect gut • Microbial diversity • Total heterotrophic bacterial populations (THBP) • Hydrolytic extracellular enzyme

4.1 Microbial Diversity in Insect Gut

Studies of associations between artificial diet feeding and gut microorganisms of selected insects have sharpened appreciation for utilizing the diet in predator rearing gained importance. In addition to enriching the fundamental ecological knowledge, also bring the advancements in applications as diverse as biological control of insect pests. In this study, it tested the hypotheses that host artificial diet has any influence on microbial community composition in *Rhynocoris marginatus* adults. The composition of microbes as well as their population differs when nutrient diet sources are varied. Further, microbes' hydrolytic enzyme level also differs in the fore-, mid-, and hindgut.

Studies of insect–microbe interactions have produced advancements in the applications as diverse as biological control of insect pests. Fragment length polymorphisms (FLP) analyses performed to identify the gut microbes. Due to the nonavailability of the facility is used to routine methods. It is simple, affordable, and reliable, and many scientists have been still following the same in their researches. The insect gut is inhabited by a wide diversity of microorganisms as a result of continuous exposure to the external and internal environments. Diet compositions place an important role in governing the gut microbial populations and their activities. Physiological (e.g., host-searching ability, enzyme, pheromone, and sound production) and laboratory rearing could induce changes and affect the physiological state of insects; in turn, it will affect their performance. For example, host-searching ability of the parasite may be reduced, or the ability to produce pheromone, enzymes, sound, etc., is impaired. In some cases visual perception is affected. Singh (1982) has given examples of how these factors can affect insects in culture. This physiological factors should be thoroughly studied if the rearing has to be maintained over several generations.

The diversity of the Insecta is reflected in the large and varied microbial communities inhabiting the gut (Dillon and Dillon 2004a, b). The indigenous gut bacteria is regarded as a valuable metabolic resource to the nutrition of the host by improving the ability to live on suboptimal diets, improved digestion efficiency, acquisition of digestive enzymes, and provision of vitamins (Douglas 1992; Tanada and Kaya 1993; Breznak and Brune 1994; Biggs and Mc Grego 1994; Bignell et al. 1997; Brauman ct al. 2001; Broderick et al. 2004). The contribution of gut microbiota to nutrition and disease suppression was also studied by Dillon and Charnley (1986, 1988, 1996). Gut microbes of ground beetles *Poecilus chalcites* (Lehman et al. 2009), *Harpalus pensylvanicus* (DeGeer), *Anisodactylus sanctaecrucis* (F.) (Coleoptera: Carabidae) (Lundgren et al. 2007), and *Coleomegilla maculata* (Coleoptera: Coccinellidae) (Lundgren and Weber 2010) were available in the literature.

4.2 Microbial Diversity in Heteropteran Predator Gut

The symbiotic bacteria harbored in the midgut ceca of the Heteroptera caused aposymbiotic nymphs of several heteropterans to reportedly exhibit retarded growth and/or nymphal mortality (Abe et al. 1995) suggesting that the symbionts play some important roles for the host insects. Further, any zoophagous insects are associated with autochthonous microbes for their development and reproduction, i.e., sustenance. The rate of food consumption, digestion, assimilation, and conversion is also depending upon the same microbes. The microbiological nature of symbiotic bacteria has been poorly understood. Although several bacteria have been isolated from the gut of some heteropterans in general (Dasch et al. 1984) and reduviids in particular, it has scarcely been confirmed whether the isolates are identical to the predominant bacteria harbored in the midgut ceca.

Insects dependent on restricted diets, such as plant sap, especially hemolymph feeders, commonly carry symbiotic microorganisms that are thought to provide nutritional supplements for their hosts (Dasch et al. 1984). In this respect the Heteroptera are an interesting research subject for understanding the diversity and evolution of insect–microbe symbiotic associations, because a number of hemolymph feeders and predators of other arthropods are found in well-defined insect group.

Certainly, symbiotic relationships in the Heteroptera correlate reasonably with diet; symbiotic bacteria tend to be found in hemolymph feeders. Among the hemolymph-sucking groups of the Heteroptera, the family Reduviidae shows the most remarkable behavioral and anatomical arrangement for transmission of the symbiont. Large-scale food processing machinery is available for making hundreds or thousands of kilograms of diet per day that may be enquired in a mass-rearing facility (Rothrock 1996; Cohen 2000). The insectary worker is the major source of microbes in a rearing facility (Sikorowski 1984; Sikorowski and Goodwin 1985), and once the worker removed microbes from the diet production and rearing process as possible, the contamination problem can be solved.

Advancement of augmentative biological control would also greatly profit in post-rearing distributional technology, field release, and field evaluation systems. A number of arthropod natural enemies have been reared with variable success on artificial diet. Several predatory heteropterans, chryospids, reduviids, and coccinellids have been reared for consecutive generations on diets devoid of insect materials.

Macrolophus is a genus of polyphagous mirid predators commonly used in European greenhouses for the biological control of whiteflies, spider mites, thrips, aphids, and leaf miners. Microbial community of two commercially available biological control agents for the augmentative biological control of arthropod pests in European greenhouses such as *Macrolophus pygmaeus* and *Macrolophus caliginosus* (predatory stink bugs) (Machtelinckx et al. 2012). *Nesidiocoris tenuis* (Reuter) (Miridae) is an omnivorous insect used for biological control. Augmentative release and conservation of *N. tenuis* have been used for pest control in tomato crops. Caspi-Fluger et al. (2014) studied and characterized symbiont *Rickettsia* in this mirid bug.

Microorganism diversity in the digestive system of insect natural enemies (Reduviidae) has been a recent focus of attention. Nonetheless, nothing is known regarding the functional role of these microorganisms associated with reduviid predators including *Rhynocoris marginatus*. In the present chapter, microbiological aspects, such as microbial diversity, localization in vivo, and fitness effects on the host insect of the bacteria contained in the gut, were investigated.

4.3 Artificial Diet and Reduviid Gut Microbiology

Microbes or active biochemical factors such as enzymes are usually present in natural and factitious prey, and AD will be determined. While the oligidic diet optimizes growth, development, and reproduction, it did not interfere the bioefficacy of the predators. The bioefficacy can be determined and compared between laboratory and field experiments.

4.3.1 Dissection of the Predators

Adults of *Rhynocoris marginatus* were reared on *Corcyra cephalonica* (T_1), *Corcyra cephalonica* with water (T2), artificial diet (T3), artificial diets + *Corcyra cephalonica* (T4), and *Spodoptera litura* (T5) separately. Laboratory emerged adults of *Rhynocoris marginatus* are selected randomly prior to the morning feed when the gut was empty and kept at 4 °C for 15 min prior to handling to prevent regurgitation, surface sterilized with 0.1 % mercuric chloride solution for 2 min, and thrice washed with sterile distilled water. Under aseptic condition, each insect was carefully dissected by using pins, fine forceps, and razors in a dissection tray filled with sterile phosphate-buffered saline (pH, 6.9). Isolated guts were individually washed several times with fresh phosphate-buffered saline to minimize the possible microbial contamination and used for the study, and the dry weight of the gut was recorded.

4.3.2 Enumeration of Total Heterotrophic Bacterial Population (THMP) of Gut Content

In order to identify the gut microflora, the gut was homogenized with sterile insect Ringer's solution (IRS) in mortar and pestle. The homogenate was filtered through Whatman filter paper no. 1 and the pH was measured using pH meter. The filtrate was serially diluted in sterile saline, and 0.1 ml of aliquot was plated on nutrient agar (NA) and Trypticase soy agar (TSA). The seeded NA plates were incubated at 37 °C for 24–48 h, whereas the SDA plates were incubated at 28 °C for 24–72 h. Microbial colonies which appeared after the incubation period were enumerated, and the numbers of colony-forming units were expressed as dry weight of the gut.

Gut microbiota contribute to the health of their hosts. Numerous studies in the past distribution of the microbial gut community using classical techniques here are used in this study. The insects are rather uniform in their nutritional requirements; if it changes automatically, their indigenous microbial population changes. Because microbes are an integral part of the insect, they undergo change when the ecological and physiological processes change in the gut. Douglas (2009) recorded in the article that gut microbes of predatory insects are not known. Hence, information related to a predatory reduviid is highlighted here.

4.4 Total Heterotrophic Bacterial Population

The total heterotrophic bacterial population is varied in the diet ingestion by the reduviid *Rhynocoris marginatus*. Among all the categories, the bacterial population was found to be higher in *Spodoptera litura* (114.32) category followed by *Corcyra cephalonica* (103.8), *Corcyra cephalonica* with water (98.55), and oligidic diet with *Corcyra cephalonica* (68.5). THBP was drastically decreased in artificial diet-reared reduviids (40.4) (Fig. 4.1). It showed that microbe's population was drastically altered if the food nature changed.

Natural food like *Spodoptera litura* keeps the gut microbes viable and enriches gut natural population. However, apart from *Spodoptera litura*, the microbial colonization in oligidic diet with *Corcyra cephalonica* group clearly shows that the indigenous microbial community has reduced due to artificial diet feeding. This should be enriched by modulating other ingredients in the artificial diet. It was reported by Dillon and Dillon (2004a, b) that many species derive their microbiota from the surrounding environment such as the phylloplane of food plants or the skin of the animal host, but the degree of persistence of strains of the ingested species is unknown. This postulation supports our findings too:

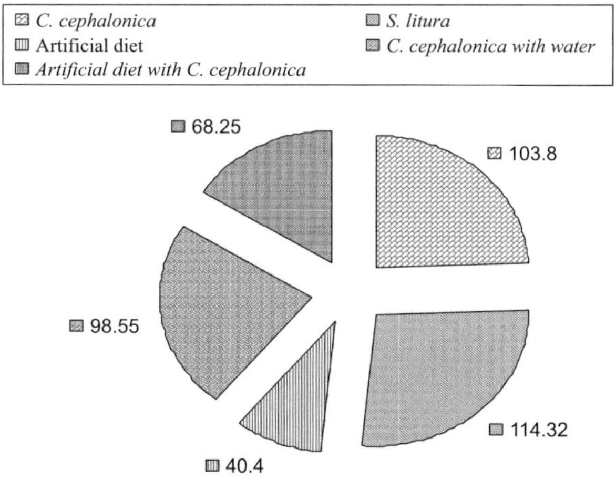

Fig. 4.1 Mean total viable bacterial count (CFU × 10^3 ml/g) of *Rhynocoris marginatus* reared on artificial diet and host

1. *Reduction of microbial species number*
2. *Reduction of particular species in different diets*
3. *Reduction of total number of species as well as total number of bacterial populations*

Latter the idea was also supported by Douglas (2009) who reported that many insects derive nutritional advantage from persistent associations with microorganisms that variously synthesize essential nutrients or digest and detoxify ingested food. These persistent relationships are symbioses he added.

4.4.1 Identification of Bacteria

Trypticase soy agar, MacConkey agar, pseudogel, polymyxin private egg yolk, mannitol, and bromothymol blue agar were used for isolating total bacteria, enterics, pseudomonas, and *Bacillus* species (Hunt and Charnley 1981). Different morphological microbial colonies were selected, subcultured, and stored at 4 °C on respective agar slants. Bacterial strains were identified using the criteria suggested by Buchanon and Gibbons (1979) based on morphological, cultural, and biochemical characteristics like gram staining, morphology, motility, catalase, indole, methyl red, Voges–Proskauer, citrate utilization, starch hydrolysis, gelatin hydrolysis, spore test, glucose, arabinose, xylose, lactose, sucrose, raffinose, galactose, maltose, mannitol, oxidase tests, etc.

Bacteria are a large group of single-celled prokaryote microorganisms and are ubiquitous in every habitat on Earth, growing in soil, in water, as well as in organic matter, plants, and animals. A total of 650 isolates of bacteria were grouped based on their morphological characters and their ability to assimilate different types of carbohydrates and urea. Further characterization with fermentation tests suggested that they belong to eight genera like *Bacillus*, *Carnobacterium*, *Escherichia*, *Klebsiella*, *Micrococcus*, *Proteus*, *Staphylococcus*, and *Streptococcus* which are consistently associated with animal hosts and frequently observed in clone libraries from arthropod digestive tracts. Though expect more genera from the reduviids, the laboratory rearing imposed a different and less diverse diet on the reduviid than in the field.

In aphids, the genera *Enterobacter*, *Serratia*, *Klebsiella*, and *Erwinia* were the common gut bacteria (Harada et al. 1996). Fourteen bacterial species such as *Bacillus subtilis*, *Bacillus megaterium*, *Bacillus cereus*, *Corynebacterium xerosis*, *Corynebacterium kutscheri*, *Escherichia coli*, *Klebsiella pneumonia*, *Pseudomonas aeruginosa*, *Micrococcus varians*, *Micrococcus luteus*, *Proteus vulgaris*, *Staphylococcus aureus*, *Staphylococcus epidermidis*, and *Staphylococcus saprophyticus* belong to eight genera and were found to be common in the gut of *Rhynocoris marginatus* reared in three different diet regimes.

All bacterial species (14) were recorded in artificial diet-reared *Rhynocoris marginatus*, followed by the *Corcyra cephalonica* (9) and *Spodoptera litura* (9). *Micrococcus* sp., *Bacillus subtilis*, *Pseudomonas fluorescens*, *Streptococcus faecalis*, and *Aeromonas* sp. were also isolated from the predators. *Micrococcus varians* was a dominant (52.70 %) species diet. A biological role of recorded microbes was available in the literature worldwide. We presented the same in a nutshell here. *Micrococcus varians* in water and soil has a role of oil degradation (Khan and Singh 2011; Omotayo et al. 2012). But recent studies show that protease activity (Kumari 2014).

4.5 Gut Autochthonous Microbes

The gut microbiota of insects plays crucial roles in the growth, development, and environmental adaptation to the host insects or food provided. While rear the insect with artificial diet, it is necessary to record the gut microbes and their activity. The results mentioned below indicate that artificial diet molecules have the capacity to change the bacterial population which change the digesting and relevant physiological activities.

4.5.1 *Bacillus* spp.

Bacillus cereus are large, rod-shaped, aerobic, endospore-forming, gram-positive bacteria which are members of the genus *Bacillus* existing in laboratory cultures worldwide. *Bacillus cereus* is widespread in nature and frequently isolated from soil and growing plants, but it is also well adapted for growth in the intestinal tract of insects and mammals. *Bacillus subtilis* showed chitinase activity (Chandrasekaran et al. 2012; Senol et al. 2014), also by utilizing amino sugars, and is involved in glucose catabolism says Gaugué and co-workers (2013). The strong fibrin-specific enzyme, namely, fibrinolytic enzyme, was also isolated from this bacteria which had a high degrading activity for the β-chains and α-chain of fibrin(ogen) (Wang et al. 2006).

Bacillus megaterium is a rod-like, gram-positive, mainly aerobic spore-forming bacterium found in widely diverse habitats. With a cell length of up to 4 μm and a diameter of 1.5 μm, *Bacillus megaterium* is among the biggest known bacteria. The cells often occur in pairs and chains, where the cells are joined together by polysaccharides on the cell walls. *Bacillus megaterium* has been recognized as an endophyte and is a potential agent for the biocontrol of plant diseases. It produces enzymes for modifying corticosteroids, as well as several amino acid dehydrogenases. It was identified from the gut of many insects like Pacific Coast wireworm, *Limonius canus* LeConte, a serious pest of potato (Lacey et al. 2007). It supports the production of amylase an important hydrolytic enzyme (Prasanna et al. 2014) responsible for digestion of carbohydrate digestion.

Bacillus cereus cells produce several extracellular degradative enzymes such as phospholipase C, enterotoxins, and hemolysins, which are putative virulence factors. The exact role of *Bacillus megaterium* is now known well in the literature; however, it is used as protoplast. *Bacillus cereus* is a spore-forming gram-positive bacterium and also includes the closely related insect pathogen *Bacillus thuringiensis*. *Bacillus cereus* is an intestinal symbiont of animals including insects (Margulis

Table 4.1 Viable bacterial count of *Rhynocoris marginatus* reared on *Corcyra cephalonica* (T_1), *Corcyra cephalonica* with water (T2), artificial diet (T3), artificial diets + *Corcyra cephalonica* (T4), and *Spodoptera litura* (T5)

Proportion of bacterial population					
Bacteria	T1	T2	T3	T4	T5
Bacillus subtilis	18.96	16.56	8.9	10.83	10.1
Bacillus megaterium	–	3.35	10.81	–	10.10
Bacillus cereus	8.9	8.9	7.20	–	6.1
Corynebacterium xerosis	0.92	3.55	0.71	2.18	2.40
Corynebacterium kutscheri	–	–	1.01	–	–
Escherichia coli	–	–	10.47	4.25	1.6
Klebsiella pneumoniae	1.10	2.25	2.81	–	–
Pseudomonas aeruginosa	1.14	2.25	0.5	0.55	1.4
Micrococcus varians	42.33	36.45	27.20	52.70	40.6
Micrococcus luteus	4.3	8.5	3.6	1.74	10.4
Proteus vulgaris	2.40	2.40	3.25	–	–
Staphylococcus aureus	20.33	21.35	16.11	28.0	16.56
Staphylococcus epidermidis	–	–	6.08	–	–
Staphylococcus saprophyticus	–	–	5.60	–	–

et al. 1998). It also considered as pathogen of insects larch sawfly *Pristiphora erichsonii* (Htg.) and their respective abilities to produce lecithinase (Heimpel 1955).

4.5.2 *Corynebacterium* spp.

Corynebacterium is a genus of gram-positive, catalase-positive, non-spore-forming, nonmotile, aerobic, rod-shaped, straight or slightly curved bacteria. They are widely distributed in nature and are mostly innocuous.

Corynebacterium xerosis is also a gram-positive bacilli bacterium which catalyzes the synthesis of polymetaphosphate (PMP) from ATP (Muhammed 1961). Similarly, previously, an enzyme which splits high molecular weight polymetaphosphate to orthophosphate has been extracted from *Corynebacterium xerosis* (Muhammed Amir et al. 1959). This bacteria was isolated from the gut of a detritus feeder *Gryllotalpa africana* (Beauvois) and also from pest *Oryctes monoceros* adults (Desai and Bhamre 2012a). In natural prey-reared reduviid, the bacteria were maximum in oligidic diet. In contrast, the number of bacteria within oligidic diet-reared insect decreased (Table 4.1).

4.5.3 *Escherichia coli*

Escherichia coli bacteria normally live in the intestines of people and animals. Most *Escherichia coli* are harmless and an important part of a healthy human intestinal tract. However, some *Escherichia coli* are pathogenic, they can cause illness, either diarrhea or illness outside of the intestinal tract. The types of *Escherichia coli* can cause diarrhea that can be transmitted through contaminated water or food, or through contact with animals or persons. *Escherichia coli* consist of a diverse group of bacteria.

Hemocytes of the insect food during natural predation, lipopolysaccharide (LPS) of gram-negative bacteria facilitate binding and internalization of either cell-associated or cell-free LPS (Charalambidis et al. 1996). Later it was identified as a gram-negative *Escherichia coli* (*Enterobacteriaceae*) phagocytosis. Phagocytosis is a process that involves binding and internalization of pathogens, and it is essential for host defense in higher eukaryotes like insects. Though *Escherichia coli* dwelled in the gut of *Bombyx mori*, it is considered as contaminant (Anand et al. 2010). Because *E. coli* did not utilize any of the polysaccharide substrates used, cellulose, xylan, pectin, and starch are added.

4.5.4 *Klebsiella pneumoniae*

Klebsiella pneumoniae (*Enterobacteriaceae*) is a gram-positive, nonmotile, encapsulated, lactose-fermenting, facultative anaerobic, rod-shaped bacterium. It easily degrades starch while in phytophagous gut and possesses cellulolytic activity. Previously *Klebsiella pneumoniae* was isolated from a predator *Rhizophagus grandis* (Yaman et al. 2010). It was also recorded from the gut of adult desert locusts, *Schistocerca gregaria* (Dillon et al. 2010), and life stages of the Mediterranean fruit fly (Diptera: Tephritidae) (Lauzon et al. 2009).

The bacteria was also recorded from the gut of adult *Anastrepha ludens* (Diptera: Tephritidae) (Leroy et al. 2011). Now it recorded the *Rhynocoris marginatus*. In addition, it was also reported by Ramesh et al. (2009) that it is important to consider other biologically active metabolites produced by insect gut microorganisms may also confer a protective effect on a host insect species. *Klebsiella pneumoniae*, a common gut bacterium, was involved in the degradation of endosulfan which degrades the insecticide without the formation of toxic metabolite endosulfan sulfate.

4.5.5 *Pseudomonas aeruginosa* (*Proteobacteria*)

Pseudomonas aeruginosa is a common gram-negative, aerobic, *Bacillus* bacterium that can cause disease in plants and animals, including humans. It is citrate, catalase, and oxidase positive. *Pseudomonas aeruginosa* is the chief cause of disease in many insect species and also present in vectors like *Anopheles* spp., *Rhodnius* sp. However, dwelled in the gut of many insects like desert locust *Schistocerca gregaria* (Dillon and Charnley 2002), silkworm (Anand et al. 2010), adults and larvae of *Apis mellifera carnica* (honey bee), and *Bombus terrestris* (bumble bee), as well as larvae of *Osmia bicornis* (red mason bee) (Mohr and Tebbe 2006), midgut of three Brazilian

Lutzomyia longipalpis (Lutz and Neiva) populations (Gouveia et al. 2008). First *Pseudomonas aeruginosa* observed from a reduviid predator.

4.5.6 *Micrococcus luteus*

Micrococcus luteus is a gram-positive, to gram-variable, nonmotile, coccus, saprotrophic bacterium that belongs to the family Micrococcaceae. It is urease and catalase positive. An obligate aerobe and *Micrococcus* are a common indigenous biota of insect guts.

The date palm root borer *Oryctes agamemnon* (Coleoptera: Scarabaeidae) gut has *Micrococcus varians*, *Micrococcus kristinae*, *Micrococcus roseus*, and *Micrococcus lylae* (Thakur et al. 2015). It reported to *Gryllotalpa africana* Beau. (Orthoptera: Gryllotalpidae) (Desai and Bhamre 2012b), cotton leafworm *Spodoptera littoralis* (Shao et al. 2014), *Plagiodera versicolora* (Laicharting) (Coleoptera: Chrysomelidae) (Demirci et al. 2013). Peroxiredoxins (Prxs) are antioxidative enzymes that may eliminate reactive oxygen species (ROSP) efficiently. The presence of *Micrococcus luteus* in the pea aphid *Acyrthosiphon pisum*-induced H_2O_2 production (Zhang and Lu 2015) was also recorded in the reduviid. However, the role of *Micrococcus luteus* in *Rhynocoris marginatus* was not known.

4.5.7 *Proteus vulgaris* (*Proteobacteria*: *Enterobacteriaceae*)

Proteus vulgaris is a rod-shaped, nitrate-reducing, indole + and catalase-positive, hydrogen sulfide-producing, gram-negative bacterium that inhabits the intestinal tracts of humans and animals. It can be found in soil, water, insect gut, and fecal matter. *Proteus vulgaris* contains an enzyme which catalyzes the oxidation of some 11 amino acids to their corresponding keto acids with liberation of ammonia. It was recorded from the hindgut of the cockroach *Blatta orientalis* along with *Klebsiella*, *Serratia marcescens*, *Pseudomonas aeruginosa*, and *Bacillus* spp. (Burgess et al. 1973). It was also reported from the gut of *Bombyx mori* (Anand et al. 2010), *Trioza magnoliae* (Spaulding and von Dohlen 1998), and *Aphis pisum* (Moran et al. 2005).

4.5.8 *Staphylococcus* spp.

Staphylococcus aureus, *Staphylococcus epidermidis*, and *Staphylococcus saprophyticus* were identified from *Rhynocoris marginatus* gut. *Staphylococcus aureus* is a gram-positive coccal bacterium that is a member of the *Firmicutes* and is frequently found in the respiratory tract and on the skin. It is often positive for catalase and nitrate reduction. *Staphylococcus aureus* was the most widely distributed bacteria in the gut, while *Klebsiella* spp. were the least (Ademolu and Idowu 2011) in grasshopper *Zonocerus variegatus* (Orthoptera: Pyrgomorphidae).

Staphylococcus epidermidis is a gram-positive bacterium and one of over 40 species belonging to the genus *Staphylococcus*. It is part of the normal human flora, typically the skin flora, and less commonly the mucosal flora. Although *Staphylococcus epidermidis* is not usually pathogenic, patients with compromised immune systems are at risk of developing infection. Previously three culturable bacteria, namely, *Bacillus subtilis*, *Staphylococcus gallinarum*, and *Staphylococcus saprophyticus*, were isolated from mealy bug, *Rhizoecus amorphophalli* (Ravikumar Sreekala Sreerag et al. 2014). *Staphylococcus saprophyticus* is a gram-positive, coagulase-negative bacterium belonging to the *Staphylococcus* genus. *Staphylococcus saprophyticus* is a common cause of community-acquired urinary tract infections.

From *Aleurocanthus woglumi* Ashby (Hemiptera: Aleyrodidae), Pandey and co-workers (2013) isolated 30 bacterial phylotypes belonging to 14 different genera, namely, *Bacillus*, *Kocuria*, *Micrococcus*, *Staphylococcus*, *Paenibacillus*, *Rhodococcus*, *Rummellibacillus*, *Arthrobacter*, *Curtobacterium*, *Psychrobacillus*, *Listeria*, *Brevibacillus*, *Bhargavae*, and *Pantoea*, by culturable methods as we recorded in *Rhynocoris marginatus*. Sahayaraj (2007) first isolated and identified the gut bacteria of three reduviids such as *Acanthaspis pedestris* Stall, *Haematorrhophus nigroriolacous* (Router), and *Catamearus brevepennis* Distant. The gut microflora represents all the aspects of microbial relationship from pathogenic to obligate mutualism (Dillon and Dillon 2004a, b; Dillon and Charnley 1996). In natural prey-reared reduviid, the number of bacterial cells remained almost constant during the experiment, while the number of bacteria than other artificial diet increased. In contrast, the number of bacterial species within artificial diet-reared insect decreased. In natural prey, *Spodoptera litura*, nine bacteria were observed with varying number of bacterial cells. In T1 categories, ten bacteria species were isolated followed by in T3 categories (artificial diet). Fourteen bacterial species were isolated from T4 category (artificial diets + *Corcyra cephalonica*).

The following bacteria were found in *Rhynocoris marginatus* adult: *Micrococcus* spp., *Bacillus* sp., *P. aeruginosa*, *C. xerosis*, and *S. aureus*. They are found in the three categories suggesting that they were not transient to true intestinal bacteria of this reduviid. Brooks (1963), Tanada and Kaya (1993) and Takatsuka and Kunimi (2002) reported that *Staphylococcus* spp. are the typical intestinal bacteria found in insects. It was also reported that *Klebsiella* spp. are the autochthonous bacteria of insects. In addition, we recorded *Bacillus* spp. and *Carnobacterium* spp., were the common bacteria particularly in hemipteran bug. Dillon and Charnley (1995) identified *Bacillus* spp., *E. coli, and Micrococcus* spp., from the fecal extracts of desert locust *Schistocerca gregaria*.

Among the observed bacteria, *M. varians* was the dominant species, its proportion was ranged from 27.20 to 52.70 %. Natural host *Spodoptera litura*-reared *Rhynocoris marginatus* has only a bacterial species, which increased to artificial diet-reared predator. In artificial diet, common bacteria *Escherichia coli* dominated followed by *Staphylococcus epidermidis* and *Staphylococcus saprophyticus*. This leaves three options, to overcome the presence of various nutrients and chemicals impacts predator produced diversified bacteria species, bacterial diversity probably in directly correlated with reduviid feeding which suggests that the reduviid do not have control over bacterial diversity.

In adult insects, on the other hand, the symbiont shows a specific localization to the posterior section of the midgut. Therefore, the symbiont must colonize the specific part of the midgut in the developmental course of *Rhynocoris marginatus*. In adult insects, notably, the alimentary tract is segmented and differentiated into a series of peculiarly compartmentalized structures. It will be of great interest to investigate the development of the alimentary canal of *Rhynocoris marginatus* in connection with the localization of the symbiont such as *Bacillus* spp., *Carnobacterium* spp., and *Micrococcus* spp., are the common bacteria in all the diets.

The bacteria found in *Rhynocoris marginatus* were *Streptococcus* spp. and *Staphylococcus* spp., which are typical intestinal bacteria found in insects (Brooks 1963; Tanada and Kaya 1993). In adult insects bacterium showed a specific localization to the posterior section of the midgut. Disrupted transmission of the bacterium to newborn nymph diet or eggs or saliva is deposited by other reduviids, and these had to grow inside the egg of the insect resulting in retarded development, arrested growth, abnormal body coloration, and other symptoms. From these results we concluded that the bacterium is a mutualistic gut symbiont of *Rhynocoris marginatus* which is vertically transmitted through the egg capsule or food and is essential for normal development and growth of the host insect.

The diversity of symbiotic bacteria as well as endosymbiotic mechanisms is the Heteroptera. It suggests that the endosymbiotic associations have evolved multiple times independently in a dynamic manner, although the number of species examined is quite limited. Examination of more species from diverse taxa of the Heteroptera is needed. Notably, the insects with a trace amount of the symbiont also exhibited the deficient symptoms, suggesting the possibility that a threshold titer of the symbiont must be ingested by newborn nymphs to establish normal endosymbiotic association. The symbionts probably provide the host insect with nutritional supplements, such as essential amino acids and vitamins, and have been reported for other plant-sucking insects (Baumann and Maran 1997; Douglas 1992).

4.6 Hydrolytic Extracellular Enzyme

The artificial rearing of predatory insects started a long time ago, with the main goal to try obtaining a mean to multiply and produce predatory to be released in biological control strategies. But it is also a powerful tool to conduct studies on biology, physiology, and behavior of entomophagous, especially entamopredator, species. Furthermore, for a complete success in rearing entomophagous insects are artificial conditions, all their physiological requirements have to be fulfilled, but nutrition is one of the most critical functions. Generally speaking, any nonnatural food used to produce an insect for laboratory studies could be well known in composition to draw reliable conclusions of the achieved tests. Hence, in the forth-coming part of this chapter, how natural as well as laboratory host and artificial diet influence the enzyme profile of this reduviid has been discussed.

The extracellular enzymes like amylase, protease, cellulase, and gelatinase were tested by using the nutrient media containing 0.2 % (w/v) carboxymethyl cellulose (cellulase), starch (amylase), and skimmed milk powder (protease) as substrates. Pure culture of each bacterium was isolated and streaked on respective media, and utilization of these substrates was determined by observing the clear zone around the colonies (Buchanon and Gibbons 1979). The screening experiments were performed in triplicates.

A total of 650 bacterial isolates produce amylase, protease, and lipase (P + A) (Fig. 4.2). Irrespective of the prey categories from T1 to T5 [*Corcyra cephalonica* (T1), *Corcyra cephalonica* with water (T2), artificial diet (T3), artificial diets + *Corcyra cephalonica* (T4), and *Spodoptera litura* (T5)], the amylolytic activity was observed. In AD category amylase, protease, and invertase activities were observed in the foregut, and the activities were reduced in the hindgut (except protease) and salivary gland (except amylase). Hydrolytic activity was lower in factitious host-reared predators. Lipase and invertase activities were very high in the control category. Amylase, invertase, protease, and lipase activities were higher in the foregut and hindgut (except amylase) and salivary gland of the control categories.

The digestive tract of insects is divided into three main regions: foregut and hindgut, which are derived from the ectoderm, and the midgut, the main organ where digestion and absorption take place, which has an endodermic origin. Digestion is the process by which large size food molecules are broken down into smaller molecules that are able to be absorbed by the gut tissue. This process is controlled by digestive enzymes and depends on their location on the insect gut. The initial digestion of proteins starts with the action of proteinases (endopeptidases), the intermediate digestion carried out by the exopeptidases, and the final digestion by dipeptidases. In carbohydrates, the initial and intermediate digestion is performed by alpha-amylase and the final by alpha-glucosidase. Most food molecules requiring digestion are polymers, such as proteins and starch (or glycogen), and are subsequently digested through three phases:

1. *Primary digestion is the dispersion and reduction in molecular size of the polymers and results in oligomers.*
2. *During intermediate digestion, these undergo a further reduction in molecular size to dimers.*
3. *In final digestion dimers are converted into monomers.*

The different phases of digestion occur at different compartments inside the midgut. In the case of insects having a peritrophic membrane (PM), initial digestion occurs inside PM, the intermediate digestion outside PM, and final digestion at the surface midgut cells carried out by membrane-bound enzymes. Compartmentalization of digestion increases the efficiency of the digestive process. In the case of insects lacking a PM, as exemplified by hemipterans, the midgut microvillar membranes are ensheathed an unusual extracellular lipoprotein membrane. This membrane was

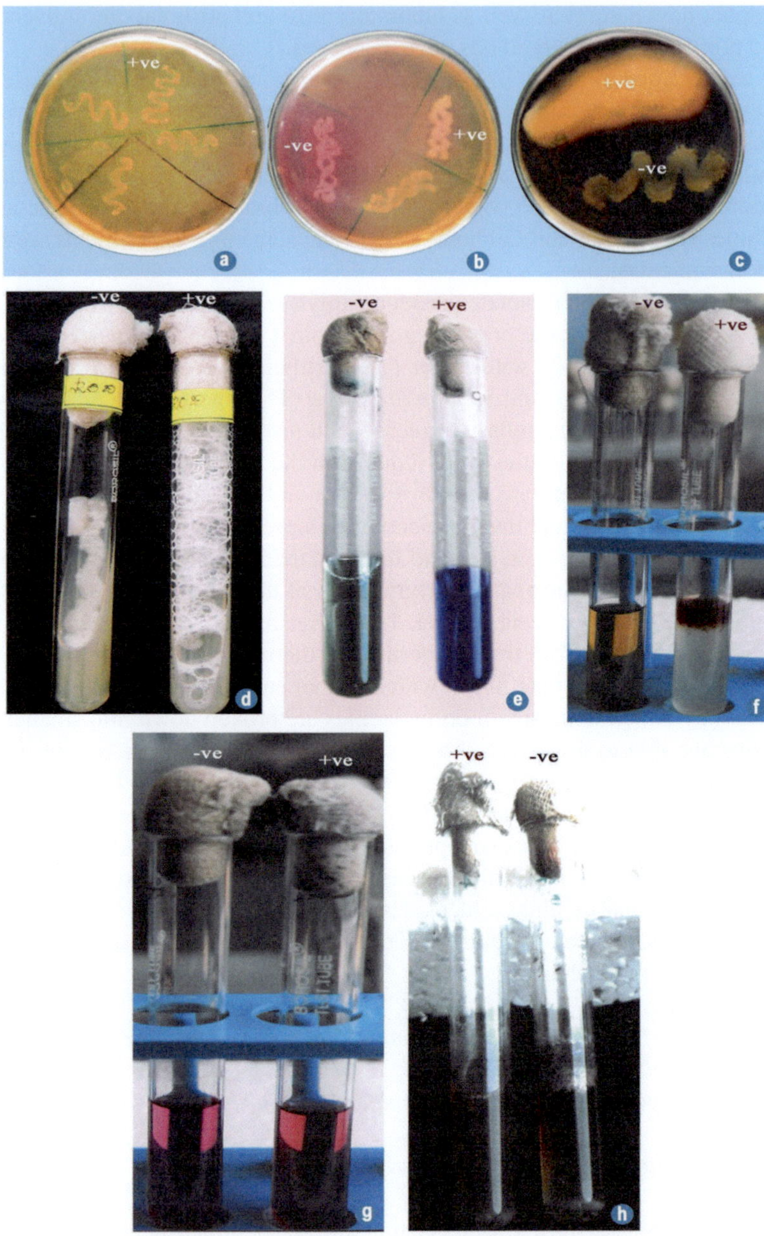

Fig. 4.2 Lactose (**a**, **b**), carbohydrate (**e**) fementation, amylase (**c**), catalase (**d**), methyle red (**e**), voges proskar (**g**) and nitrate reduction test (**h**) of reduviid predator gut microbes

named perimicrovillar membrane (PMv) and is widespread among paraneopteran insects. PMv limits a closed space, the perimicrovillar space, and in hemipterans, digestion occurs into the lumen and perimicrovillar space at the microvillar membrane surface.

4.7 Conclusions

The study of the microbiota in the reduviid predator gut has gained increased attention because of the possible role it may play in the predation on natural, laboratory, and artificial diets. Traditional methods (colony-forming unit, CFU, morphological and biochemical) have been used for the identification and enzyme profiling of the isolated microbes. More number of microbes dwelled in the gut of reduviid when red with natural hosts, and altered it change the diet to laboratory host (*Corcyra cephalonica*) and artificial diet. Provision of water along with insect food also changes the gut microbial populations.

4.8 Future Recommendations

- Identify pathogenic virulent microbes and utilize them for biological control.
- Very accurate methods like polymerase chain reaction (PCR), denaturing gradient gel electrophoresis (DGGE), real-time polymerase chain reaction (RT-PCR), and analysis of the phenol oxidase (PO) activity can be useful for the correct identification of indigenous microbes.
- Rare and specific microbes can be utilized for peptide synthesis/anti-compounds and also insect control purposes.

References

Abe Y, Mishiro K, Takanashi M (1995) Symbiont of brown-winged green bug, *Plautia stali* Scott. J Appl Entomol Zool 39:105–115

Ademolu KO, Idowu AB (2011) Occurrence and distribution of microflora in the gut regions of the variegated grasshopper *Zonocerus variegatus* (Orthoptera: Pyrgomorphidae) during development. Zool Stud 50(4):409–415

Amir M, Rodgers A, Hughes DE (1959) Purification and properties of a polymetaphosphatase from *Corynebacterium xerosis*. J Gen Microbiol 20(3):482–495

Anand AA, Vennison SJ, Sankar SG, Prabhu DI, Vasan PT, Raghuraman T, Geoffrey CJ, Vendan SE (2010) Isolation and characterization of bacteria from the gut of *Bombyx mori* that degrade cellulose, xylan, pectin and starch and their impact on digestion. J Insect Sci 10:107. doi:10.1673/031.010.10701

Baumann P, Moran NA (1997) Non-cultivable microorganisms from symbiotic association of insect and other hosts. Anton Leeuw 72:38–48

Biggs DR, Mc Greego PG (1994) Gut pH and amylase and protease activity in larvae of the New Zealand grues grab (*Costelytra zealanticus*) (Coleptera: Scarbaeidae) as a basis for selecting inhibitors. Insect Biochem Mol Biol 26:69–75

Bignell DE, Eggleton P, Nunes L, Thomas KL (1997) Termites as mediators of carbon fluxes in tropical forest budgets for carbon dioxide and methane emission. In: Watt AD, Stork NE, Hunter MD (eds) Forest and insects. Chapman and Hall Publication, London, pp 109–234

Brauman A, Bignell DE, Tayasu I (2001) Soil feeding termites biology, microbial association and digestive mechanisms. In: Abe T, Bignell DE, Higashi M (eds) Termites evolution, sociality, symbiosis, ecology. Klower Academic publishers, Dordrecht, p 259

Breznak JA, Brune A (1994) Role of microorganisms in the digestion of lignocellulose by termines. Annu Rev Entomol 39:453–487

Broderick NA, Raffa KF, Goodman RM, Handelsman J (2004) Census of the bacterial community of the gypsy moth larval midgut by using culturing and culture independent methods. Appl Environ Microbiol 70(1):293–300

Brooks MA (1963) The microorganisms of healthy insects. In: Steinhaus EA (ed) Insect pathology- an advanced treatise. Academic, London, p 250

Buchanon RE, Gibbons NE (1979) Bergey's manual of determinative bacteriology, 8th edn. Williams and Wilkins, Baltimore, p 269

Burgess NRH, McDermott SN, Whiting J (1973) Aerobic bacteria occurring in the hind-gut of the cockroach, Blatta orientalis. J Hyg 71(01):1–8

Caspi-Fluger A, Inbar M, Steinberg S, Friedmann Y, Freund M, Mozes-Daube N, Zchori-Fein E (2014) Characterization of the symbiont rickettsia in the mirid bug *Nesidiocoris tenuis* (Reuter) (Heteroptera: Miridae). Bull Entomol Res 104(06):681–688

Chandrasekaran R, Revathi K, Nisha S, Kirubakaran SA, Sathish-Narayanan S, Senthil-Nathan S (2012) Physiological effect of chitinase purified from *Bacillus subtilis* against the tobacco cutworm spodoptera litura Fab. Pestic Biochem Physiol 104(1):65–71

Charalambidis ND, Foukas LC, Marmaras VJ (1996) Covalent association of lipopolysaccharide at the hemocyte surface of insects is an initial step for its internalization. Eur J Biochem 236(1):200–206

Cohen AC (2000) Feeding fitness and quality of domesticated and feral predators: effects of long-term rearing on artificial diet. Biol Control 13:49–54

Dasch GA, Weiss F, Chang KP (1984) Endosymbionts of insects. In: Krieg NR, Holt JG (eds) Bengey's manual of systemic bacteriology. Williams and Wilkins, Baltimore, pp 811–813

Demirci M, Sevim E, Demir I, Sevim A (2013) Culturable bacterial microbiota of *Plagiodera versicolora* (L.) (Coleoptera: Chrysomelidae) and virulence of the isolated strains. Folia Microbiol 58(3):201–210

Desai A, Bhamre P (2012a) Diversity of gut bacterial fauna *of Oryctes monocerus L* innaeus (Coleoptera: Scarabaeidae). Bionano Front 5:1–4

Desai AE, Bhamre PR (2012b) Novel gut bacterial fauna of *Gryllotalpa africana* Beau. (Orthoptera: Gryllotalpidae). Int J Life Sci 6(1):50–55

Dillon RJ, Charnley AK (1986) Inhibition of *Metarhizium anisopliae* by the bacterial flora of the desert locust *Schistocerca gergari*: evidence for an antifungal toxin. J Invertebr Pathol 47:300–310

Dillon RJ, Charnley AK (1988) Inhibition of *Metarhizium anisopilae* by the gut bacterial flora of the desert locust- characterization of antifungal toxins. Can J Microbiol 34:1075–1082

Dillon RJ, Charnley AK (1995) Chemical barriers to gut infection in the desert locust: in vivo production of antimicrobial phenols associated with the bacterium *Pantoea agglomerans*. J Invertebr Pathol 66(1):72–75

Dillon RJ, Charnley AK (1996) Colonization of the gut of germ free desert locust, *Schistocerca gregaria* by the bacterium *Pantoea agglomerans*. J Invertebr Pathol 67:11–14

Dillon R, Charnley K (2002) Mutualism between the desert locust *Schistocerca gregaria* and its gut microbiota. Res Microbiol 153(8):503–509

Dillon RJ, Dillon VM (2004a) The gut bacteria of insects: nonpathogenic interactions. Annu Rev Entomol 49(1):71–92

Dillon RJ, Dillon VM (2004b) The gut bacteria of insects: non-pathogenic interactions. Annu Rev Entomol 49:1–16

Dillon RJ, Webster G, Weightman AJ, Charnley AK (2010) Diversity of gut microbiota increases with aging and starvation in the desert locust. Antonie Van Leeuwenhoek 97(1):69–77

Douglas AE (1992) Microbial brokers of insect-plant interactions. Proceedings of 88th international symposium on insect-plant relationships. Kluwer, Dordecht, pp 329–336

Douglas AE (2009) The microbial dimension in insect nutritional ecology. Funct Ecol 23:38–47. doi:10.1111/j.1365-2435.2008.01442.x

Gaugué I, Oberto J, Putzer H, Plumbridge J (2013) The use of amino sugars by*Bacillus subtilis* : presence of a unique operon for the catabolism of glucosamine. PLoS ONE 8(5), e63025. doi:10.1371/journal.pone.0063025

Gouveia C, Asensi MD, Zahner V, Rangel EF, de Oliveira SM (2008) Study on the bacterial midgut microbiota associated to different brazilian populations of *Lutzomyia longipalpis* (Lutz and Neiva) (Diptera: Psychodidae). Neotrop Entomol 37(5):597–601

Harada H, Oyaizu H, Ishikawa H (1996) A consideration about the origin of aphid intracellular symbiont in connection with gut bacterial flora. J Gen Appl Microbiol 42(1):17–26

Heimpel AM (1955) Investigations of the mode of action of strains of Bacillus cereus Fr. and Fr. pathogenic for the larch sawfly, *Pristiphora erichsonii* (Htg.). Can J Zool 33(4):311–326

Hunt J, Charnley AK (1981) Abundance and distribution of the gut flora of the desert locust *Schistocerca gregaria*. J Invertebr Pathol 38:378–385

Khan JA, Singh S (2011) Evaluation of oil degradation potential of Micrococcus varians. Int J Appl Microbiol Pharm Technol 2:75–80

Kumari S (2014) Extracellular Protease Enzyme Production using Micrococcus luteus-4, *Staphylococcus hyicus*, *Micrococcus luteus*-1, *Pasteurella pneumotrop* and *Micrococcus* sp. isolated From Water Reservoirs. Int J Curr Microbiol Appl Sci 3(5):772–784

Lacey LA, Unruh TR, Simkins H, Thomsen-Archer K (2007) Gut bacteria associated with the Pacific Coast wireworm, *Limonius canus*, inferred from 16s rDNA sequences and their implications for control. Phys Chem Chem Phys 35(5):479–489

Lauzon CR, McCombs SD, Potter SE, Peabody NC (2009) Establishment and vertical passage of *Enterobacter* (*Pantoea*) *agglomerans* and *Klebsiella pneumoniae* through all life stages of the Mediterranean fruit fly (Diptera: Tephritidae). Ann Entomol Soc Am 102(1):85–95

Lehman RM, Lundgren JG, Petzke LM (2009) Bacterial communities associated with the digestive tract of the predatory ground beetle, *Poecilus chalcites*, and their modification by laboratory rearing and antibiotic treatment. Microb Ecol 57(2):349–358

Leroy PD, Sabri A, Verheggen FJ, Francis F, Thonart P, Haubruge E (2011) The semiochemically mediated interactions between bacteria and insects. Chemoecology 21(3):113–122

Lundgren JG, Weber DC (2010) Changes in digestive rate of a predatory beetle over its larval stage: implications for dietary breadth. J Insect Physiol 56(4):431–437

Lundgren JG, Lehman RM, Chee-Sanford J (2007) Bacterial communities within digestive tracts of ground beetles (Coleoptera: Carabidae). Ann Entomol Soc Am 100(2):275–282

Machtelinckx T, Van Leeuwen T, Van De Wiele T, Boon N, De Vos WH, Sanchez JA, De Clercq P (2012) Microbial community of predatory bugs of the genus *Macrolophus* (Hemiptera: Miridae). BMC Microbiol 12(Suppl 1):S9

Margulis L, Jorgensen JZ, Dolan S, Kolchinsky R, Rainey FA, Lo SC (1998) The Arthromitus stage of *Bacillus cereus*: intestinal symbionts of animals. Proc Natl Acad Sci 95(3):1236–1241

Mohr KI, Tebbe CC (2006) Diversity and phylotype consistency of bacteria in the guts of three bee species (Apoidea) at an oilseed rape field. Environ Microbiol 8(2):258–272

Moran NA, Russell JA, Koga R, Fukatsu T (2005) Evolutionary relationships of three new species of *Enterobacteriaceae* living as symbionts of aphids and other insects. Appl Environ Microbiol 71(6):3302–3310

Muhammed A (1961) Studies in biosynthesis of polymetaphosphate by an enzyme from *Corynebacterium xerosis*. Biochim Biophys Acta 54(1):121–132

Omotayo AE, Ojo OY, Amund OO (2012) Crude oil degradation by microorganisms in soil composts. Res J Microbiol 7(4):209

Pandey N, Singh A, Rana VS, Rajagopal R (2013) Molecular characterization and analysis of bacterial diversity in *Aleurocanthus woglumi* (Hemiptera: Aleyrodidae). Environ Entomol 42(6):1257–1264

Prasanna VA, Kayalvizhi N, Rameshkumar N, Suganya T, Krishnan M (2014) Characterization of amylase producing *Bacillus megaterium* from the gut microbiota of silkworm *Bombyx mori*. Res J Chem Environ 18(7):38–45

Ramesh GK, Thangamalar A, Muthuswami M, Subramanian S (2009) Characterisation of gram negative bacterial isolates from guts of few multivoltine silkworm breeds. Karnataka J Agric Sci 22(3-Spl.):517–518

Rothrock S (1996) Food engineering's foodmaster'97, equipment, supplies, and services 1997–1997 (Section II). Foodmaster, Radnor, p 222

Sahayaraj K (2007) Isolation, identification and characterization of gut flora of three reduviid predators. Asian J Microbiol Biotechnol Environ Sci 9(4):1073–1075

Senol M, Nadaroglu H, Dikbas N, Kotan R (2014) Purification of Chitinase enzymes from *Bacillus subtilis* bacteria TV-125, investigation of kinetic properties and antifungal activity against *Fusarium culmorum*. Ann Clin Microbiol Antimicrob 13(1):1–7

Shao Y, Arias-Cordero E, Guo H, Bartram S, Boland W (2014) In vivo Pyro-SIP assessing active gut microbiota of the cotton leafworm, *Spodoptera littoralis*. PLoS ONE 9(1):e85948. doi:10.1371/journal.pone.0085948

Sikorowski PP (1984) Pathogens and microbiological contaminants: their occurrence and control. In: Sikorowski PP, Griffin JG, Roberson J, Lindig OH (eds) Boll weevil mass rearing technology. University Press of Mississippi, Jackson, pp 115–169

Sikorowski PP, Goodwin RH (1985) Contaminant control and disease recognition in laboratory cultures. In: Singh P, Moore RF (eds) Handbook of insect rearing, vol I. Elsevier, New York, pp 85–105

Singh P (1982) The rearing of beneficial insects. NZ Entomol 7(3):304–310

Spaulding AW, von Dohlen CD (1998) Phylogenetic characterization and molecular evolution of bacterial endosymbionts in psyllids (Hemiptera: Sternorrhyncha). Mol Biol Evol 15(11):1506–1513

Takatsuka J, Kunimi Y (2002) Lethal effects of *Spodoptera exigua* nucleopolyhedrovirus isolated in Shiga Prefecture, Japan, on larvae of the beet armyworm, *Spodoptera exigua* (Lepidoptera: Noctuidae). Applied Entomol Zool 37(1):93–101

Tanada Y, Kaya HK (1993) Insect pathology. Academic, London, p 276

Thakur A, Dhammi P, Saini HS, Kaur S (2015) Pathogenicity of bacteria isolated from gut of *Spodoptera litura* (Lepidoptera: Noctuidae) and fitness costs of insect associated with consumption of bacteria. J Invertebr Pathol 127:38–46

Wang CT, Ji BP, Li B, Nout R, Li PL, Ji H, Chen LF (2006) Purification and characterization of a fibrinolytic enzyme of *Bacillus subtilis* DC33, isolated from Chinese traditional Douchi. J Ind Microbiol Biotechnol 33(9):750–758

Yaman M, Ertürk Ö, Aslan I (2010) Isolation of some pathogenic bacteria from the great spruce bark beetle, *Dendroctonus micans* and its specific predator, *Rhizophagus grandis*. Folia Microbiol 55(1):35–38

Zhang Y, Lu Z (2015) Peroxiredoxin 1 protects the pea aphid *Acyrthosiphon pisum* from oxidative stress induced by *Micrococcus luteus* infection. J Invertebr Pathol 127:115–121

Gut Enzyme Profile

5

Abstract

The diverse feeding habits of *Rhynocoris marginatus* make the Reduviidae ideal for studies of feeding strategies, including digestive enzyme composition. Wide ranges of digestive enzymes were recorded in the alimentary canal of insects, and their level varies in relation to diet. Digestive enzymes play a major role in insect physiology by converting complex food materials into micromolecules necessary to provide energy and metabolites for growth, development, and other vital functions. These enzymes are produced and distributed in various regions of the gut and salivary gland in different proportions and quantities. It is hypothesized macromolecules in the prey dramatically influence reduviid growth and development mediated through qualitative and quantitative regulation of digestive enzymes. The attempt was made to study the quantitative profile of digestive enzymes in relation to different preys and oligidic diet-reared predator. *Rhynocoris marginatus* foregut and hindgut contain amylase, protease, invertase, and lipase. Amylase, invertase, and lipase activities of both the foregut and hindgut were high, while *R. marginatus* was fed with *Corcyra cephalonica* followed by artificial diet and *Spodoptera litura*. The hindgut enzyme activities were decreased (++), and similar kinds of observations were recorded for artificial diet and *Spodoptera litura*. The results show that generally prey type and artificial diet do not have any influence on the enzyme qualitative profile of this reduviid. In salivary gland, all the enzyme activities were moderate (++) except lipase (+) when *Corcyra cephalonica* and *Spodoptera litura* provided food. The activity was reduced (+) on *Corcyra cephalonica* and artificial diet for *Rhynocoris marginatus*. Results clearly show provision of artificial diet has not changed the enzyme levels in the foregut and midgut of this predator.

Keywords

Macromolecules • Salivary gland enzyme profile • Extra-oral digestion • Digestive enzymes

K. Sahayaraj, R. Balasubramanian, *Artificial Rearing of Reduviid Predators for Pest Management*, DOI 10.1007/978-981-10-2522-8_5

5.1 Enzyme Activity of Hemipteran Predators

The Hemiptera contains families: obligate zoophages (e.g., Reduviidae, Phymatinae, and Nabidae) and obligate phytophages (e.g., Tingidae and Coreidae). Further the terrestrial hemipteran predators have evolved in both the Pentatomomorpha (Geocoridae and Pentatomidae) and Cimicomorpha (Reduviidae, Miridae, Anthocoridae, and Nabidae). Several insect families contain omnivores. However, some predatory insects exclusively predate and feed on other animals (Alomar and Wiedenmann 1996). In the true bugs, enzymes are produced in specialized glands (salivary glands), pumped into the prey, and ingested into the predator's gut where they remain until the digestion completes.

The origin of feeding habits among the Heteroptera remains controversial (Sweet 1979; Cobben 1979, 1990; Schaefer 1997; Wheeler 2001). The diverse trophic habits of bugs make the Reduviidae ideal for studies of feeding strategies, including digestive enzyme composition that employs macerate (or lacerate) and flush feeding (Miles 1972; Hori 2000; Wheeler 2001) that incorporates piercing/sucking mouthparts and watery saliva from the salivary gland complex. Reduviids feed in a manner that is typical of heteropterans, piercing and cutting tissues with their stylets while injecting digestive enzymes through the salivary canal to liquefy food into nutrient rich slurry. The food slurry is ingested through the food canal and passed into the alimentary canal where it is further digested and absorbed (Cohen 2000). Digestive enzymes are produced and distributed in various regions of the gut and salivary gland in different proportions and quantities.

Wide ranges of digestive enzymes were recorded in the alimentary canal of insects, and their level varies in relation to diet. It is a well-known fact that the digestive enzymes play a major role in insect physiology by converting complex food materials into micromolecules necessary to provide energy and metabolites for growth, development, and other vital functions. A consumer's ability to use plant or animal materials for food is indicated by the presence of specific digestive enzymes (Miles 1972; Cohen 1990, 1995, 1996, 1998a, b, 2000; Agusti and Cohen 2000; Hori 2000; Zeng and Cohen 2000a, b) which include proteases, hyaluronidase lipase (Cohen 1998b, 2000), and amylase (Cohen 1996; Sahayaraj 2007). The chemical composition of the watery saliva of hemipteran insect is crucial for effective feeding because this insect relies very heavily on saliva for extra-oral digestion (Cohen 1995, 1996, 1998a). Salivary and gut enzyme compositions were studied for predatory heteropterans by Cohen (1995) and Cohen and Wheeler (1998).

5.2 Reduviid Predator Enzymology

Enzymes of reduviids were initiated by Edwards in 1961. He studied the action and composition of the saliva of an assassin bug *Platymeris rhadamanthus* Gerst. Initially the feeding behavior of *Sinea confusa* and *Zelus renardii* was studied using semisolid artificial diet (Cohen 1990). It was supported by the presence of amylase level which was high in the salivary gland of both reduviids; however, amylase level was recorded in the gut. Latter in 1993, Cohen was observed proteinases from *Zelus renardii's* salivary gland complex (SGC), anterior midgut (AMG), and posterior midgut (PMG). The digestive physiology of reduviid predators solicits greater attention in view of its economic importance. Among the digestive enzymes, amylase, invertase, lipase, and protease activities are of great importance in the digestion of food. Endopeptidase activity was found to be uniformly meager in the anterior midgut (AMG) and in the posterior midgut (PMG) in the three morphs, *viz.*:

1. *Niger (black connexivum)*
2. *Sanguineous (red connexivum)*
3. *Nigrosanguineous (black and red banded connexivum)*

Utilization of macronutrients from the available prey food depends on the digestive enzymes. Digestive enzymes of alimentary canal and the salivary gland of *Sophrorhinus insperatus* Faust (Hori 1969; Ravikumar et al. 2002) were investigated. Studies on digestive enzyme profile of Indian reduviids were not available except the preliminary works of Ambrose and Maran (2000) and Sahayaraj et al. (2007b, 2010). Prey influence on the salivary gland and gut enzyme qualitative profile of *Rhynocoris marginatus* (Fab.) and *Catamiarus brevipennis* (Serville) was recorded by Sahayaraj and his co-worker (2007b).

The nutrition needs and the knowledge of the functional organization of digestive system of reduviid predators may be useful in designing oligidic diet for mass production (says Sahayaraj et al. 2007a). Moreover, this information could be useful to understand how reduviids adopt to natural or factitious preys or artificial diets. Considering the ecological and biocontrol prospects and lacuna in this field, it investigated how the provision of natural, factitious, and artificial diets to reduviid predators affects the gut enzyme profile and salivary gland profiles.

5.3 Qualitative Enzyme Profile Methodology

The gut is the principal site for secretion of digestive enzymes, digestion of food, and absorption of nutrients. Macromolecular components present in the foods catalyzed by three major digestive enzymes in the gut, *viz.*, amylases (EC 3.2.1.1), proteases (EC 3.4.21), and lipases (EC 3.1.1.3). Amylases catalyze the hydrolysis of α-D-(1, 4)-glucan linkage in starch, glycogen, and other carbohydrates, which serve as energy sources for insects.

Gut proteases further release amino acids from peptides produced by endopeptidases. Our earlier studies indicated that changes in the gut proteases and amylases are related to host diet content. Other important but less studied enzymes from lepidopteran insects are lipases, which catalyze the hydrolysis of triacylglycerol (TAG). Lipases secreted into the midgut lumen of insects break down a variety of dietary lipids, such as triacylglycerol and phospholipids, into fatty acids. In reduviid it was observed that the products of digestion are absorbed by the midgut epithelium and

Table 5.1 Impact of *Corcyra cephalonica* (T_1), *Corcyra cephalonica* with water (T2), artificial diet (T3), artificial diets + *Corcyra cephalonica* (T4), and *Spodoptera litura* (T5) on the foregut, hindgut, and salivary gland qualitative enzyme profile of *Rhynocoris marginatus*

Treatments	Protease	Lipase	Amylase	Invertase	Trypsin	Pepsin
Foregut						
T1	+++	+++	+++	+++	++	++
T2	+++	++	++	+	+++	++
T3	+++	++	+++	+++	++	++
T4	+++	+	++	++	++	+++
T5	+++	++	++	++	+++	+++
Hindgut						
T1	+++	+++	++	+++	++	++
T2	++	++	+	+++	++	+++
T3	+++	++	++	++	++	++
T4	++	+++	++	++	+++	++
T5	+++	++	++	++	++	++
Salivary gland						
T1	+++	+	+++	+	++	+
T2	++	+	++	+	++	++
T3	++	+	+++	+	++	++
T4	+	++	++	++	++	++
T5	+++	++	+++	++	++	++

(+) less, (++) moderate, and (+++) maximum

then used to synthesize complex lipids, such as TAGs, diacylglycerols, and phospholipids.

It is hypothesized macromolecules (protein, carbohydrates, and lipids) in the prey dramatically influence reduviid growth and development mediated through qualitative and quantitative regulation of digestive enzymes. The hypothesis was tested by tracking the enzyme profile of *Rhynocoris marginatus* on five nutritionally diverse foods.

Invertase, amylase, lipase (Nigam and Omkar 2003), pepsin (Tonapi 1996), protease, and trypsin (Balogun and Fisher 1970) qualitative profiles were performed using sugar, olive oil emulsion, acid casein, peptone, casein, and alkaline casein, as substrates, respectively. Based upon the color intensity, the enzyme quantities were expressed as less (+), moderate (++), and maximum (+++) activities.

Polyphagous *Rhynocoris marginatus* feed on many preys and enjoy a better chance of survival in nature (Sahayaraj 2014). The diets comprise varying proportions of organic constituents belonging to three major groups, *viz.*, proteins, carbohydrates, and fats, which directly affect insect growth and reproduction. Food quality regulates and influences the production of alimentary canal digestive enzymes. Thus attempt was made to study the quantitative profile of digestive enzymes in relation to different preys and OD. From Table 5.1, it was very clear that *Rhynocoris marginatus* foregut and hindgut contain amylase, protease, invertase, and lipase.

But the activity profile depends on the type of diet consumed by the predators. For instance, amylase, invertase, and lipase activities of both the foregut and hindgut were high, while *Rhynocoris marginatus* was fed with *Corcyra cephalonica* followed by artificial diet and *Spodoptera litura.* The hindgut enzyme activities were decreased (++), and similar kinds of observations were recorded for artificial diet and *Spodoptera litura.* The results show that generally prey type and artificial diet do not have any influence on the enzyme qualitative profile of this reduviid. In salivary gland, all the enzyme activities were moderate (++) except lipase (+) when *Corcyra cephalonica* and *Spodoptera litura* provided food. The activity was r reduced (+) on *Corcyra cephalonica* and artificial diet for *Rhynocoris marginatus* (Table 5.1).

Results clearly show provision of artificial diet has not changed the amylase, invertase, protease, trypsin, pepsin, and lipase levels in the foregut and midgut of this predator. Rather, in the foregut, provision of artificial diet enhanced the levels of amylase, invertase, trypsin, and pepsin when compared to natural host *Spodoptera litura.* In hindgut, the levels of all observed enzymes of both artificial diet and *Spodoptera litura* fed categories are same.

The salivary glands of the reduviid contain amylase-, lipase-, and trypsin-like enzymes, which are responsible for extra-oral digestion by this predator. However, the digestion of macromolecules may also occur in the alimentary canal of this predator. The salivary glands may secrete the enzymes like protease, lipase, amylase, pepsin, invertase, and trypsin for digestion in the gut. Different diet compositions might promote specific bacterial strains by providing them with favorable nutritional conditions. Digestive enzymes play a major role by converting complex food into micromolecules which are necessary to provide energy and metabolites for growth, development, and other vital functions to the insects (Wigglesworth 1972). Food quality regulates and influences the production of alimentary canal digestive enzymes. Henceforth, an attempt was made to study the qualitative and quantitative profile of digestive enzymes in relation to different preys and artificial diet.

Results reveal that the activity profile depends on the type of diet/prey consumed by the insects. For instance, amylase, invertase, and lipase activities of both the foregut and hindgut were high, while *Rhynocoris marginatus* was fed with *Corcyra cephalonica* followed by artificial diet and *Spodoptera litura.* Conversely the true bugs have piercing and sucking mouthparts. The stylets are used to piercing the prey to disrupt the cell wall, cellular contents, and to deliver saliva containing potent digestive enzyme. The mechanical damage and the enzyme break down the tissues into slurry of small particles which are ingested along with the saliva (Cohen 1998a, b). In salivary gland, all the enzymes are moderate (++) except lipase (+) when *Rhynocoris marginatus* is reared on *Corcyra cephalonica* and *Spodoptera litura.* The results show that generally prey type and artificial diet do not have any influence on the enzyme qualitative profile of these reduviids.

Protease activity was rarely reported in gut bacterial isolates (Coolbear et al. 1992; Balasubramanian et al. 1975). Protease and amylase (P + A) activities were observed in bacterial isolates belonging to the factitious host reared *Rhynocoris marginatus*. Chemically defined diets have been developed to rear predatory insects and are highly suitable for the development of microorganisms. Various antimicrobial agents were used in insect diet to prevent microbial contamination and propagation (Yazgan and House 1970; House 1967). A limited amount of work has been done on the effect of antimicrobial agents like formaldehyde and streptomycin on predatory heteropterous species (Xie et al. 1986).

5.4 Quantitative Enzyme Bioassays

The physiology and biochemistry of the insect gut have an important role in the study of predator mass production strategies. Digestion is a phase of insect physiology on which little research has been done, considering the economic importance of the food of insects and the fact that most important control measures involve the action of digestive juices on poisons taken from the digestive tract.

Zoophagous insects, particularly generalists, face dietary nutritional variation in hosts, in terms of species, developmental stage, and tissues and responses to the environment. Usually dietary protein (DP) and digestible carbohydrates (DC) and, in particular, the ratio of these two nutrients (P:C) are critical for proper insect (e.g., *Rhynocoris marginatus*) growth and development (Simpson and Raubenheimer 1993). No previous study was available about the enzyme profile of *Rhynocoris marginatus* gut and salivary gland except the work of Sahayaraj and Balasubramanian (2008).

It is a well-known fact that the food digestion is a process in which nonabsorbable and large food molecules are broken into small molecules to be absorbed by epithelial cells of the alimentary canal. Digestion of large molecules, such as protein and carbohydrate polymers, occurs in three enzymatic phases:

1. *The first stage is conversion of polymers into oligomers.*
2. *Oligomers are cleaved into dimers during the second phase.*
3. *In the final stage, monomers are obtained.*

It was believed that three phases occur inside the insect midguts. However, it is not completely understood to which extent these stages occur inside the alimentary canal of predatory bugs like reduviids with extra-oral digestion (EOD). Further, Cohen (1995) reported that at least 79 % of terrestrial arthropods use EOD to consume food. This action allows the predators to feed on large prey and enables them to hunt insects with rigid and hard cuticles. Extra-oral digestion maximizes the efficiency of food consumption in predatory insects. The enzymes involved in EOD usually come from salivary glands or midgut and are diverse among various insect orders. Extra-oral digestion among hemipterans is reported to occur under the action of salivary enzymes. Predatory true bugs typically introduce toxins to debilitate the prey followed by digestive enzymes that liquefy the prey within its own exoskeleton. The presence of amylase has been investigated in digestive systems of many insects including members of Orthoptera, Hymenoptera, Hemiptera, Diptera,

Lepidoptera, and Coleoptera. However, in heteropteran predators, activities of amylase and proteinases (as trophic enzymes) are adjusted according to diet, and digestive enzymatic profiles can be diverse in different developmental stages.

5.4.1 Amylase

In the three important polysaccharidases in insects (amylase, cellulase, and chitinase), amylase is far best understood. In insects amylase is the most ubiquitous polysaccharidase, which digests starch to maltose, followed by a digestion via a glucosidase to glucose. Amylase is one of the key enzymes involved in digestion of carbohydrate metabolism in insects. There is no considerable literature on the amylase activity in the digestive tract of reduviids and related hemipteran predators. Hence this study will be helpful to the future reduviid biologist.

Amylase was measured using 3,5-dinitrosalicilic acid reagent prepared (DNS) according to Bernfeld (1955) and Baker (1991) methods. Take 200 μl of enzyme source in a test tube (5 ml capacity) and add substrate with buffer solutions each of 100 μl (0.2 % soluble starch in 50 Mm $CaCl_2$ and 20 Mm NaCl). The assay was terminated by the addition of 500 μl DNS. The mixture was incubated at 100 °C for 5 min, cooled after the addition of 500 μl of distilled water. The color development was read at 575 nm and composed with standard maltose hydrate. All assays were carried out at 30 °C.

To study the putative hyperproduction of amylase response of diet, the gut of reduviid adult *Rhynocoris marginatus* feeding on different preys or an artificial diet-reared reduviid gut was analyzed for amylase activity. The amylase quantity of reduviid hindgut has the higher in both preys and artificial diet than foregut. Among the five diets, artificial diet-reared reduviid hindgut has the maximum activities (0.69 mg protein released/min) followed by *Corcyra cephalonica* (0.58 mg protein released/min) and *Spodoptera litura* (0.44 mg protein released/min) (t = 48.32, df = 2, P = 0.000). Results were also observed in salivary gland, and also artificial diet showed highest activity (0.18 mg protein released/min) (t = 20.00, df = 2, P = 0.002) followed by *Corcyra cephalonica* (0.17 mg protein released/min) (t = 2.01, df = 2, P = 0.181) and *Spodoptera litura* (0.14) (t = 19.00, df = 2, P = 0.003) categories. In the present study, the foregut has the lowest amylase activity. But further increased level was recorded in the foregut of *Corcyra cephalonica* reared reduviids (Table 5.2).

High amylase activity found in the foregut alimentary canal of *Rhynocoris marginatus* suggested that amylase mainly concentrated on the foregut of the insects. This data also shows the functional role of *Rhynocoris marginatus* alimentary canal as the main zone for food digestion as observed by Cohen (1996, 1998a, b) and Agusti and Cohen (2000). They reported that *L. hesperus* saliva contains amylase activity. In the present study, amylase activity of salivary gland was great in predator reared with artificial diet than those fed with *Spodoptera litura* and *Corcyra cephalonica*. Perhaps the large amount of amylase secreted in the digestive tract of lab-reared reduviid is due to the polysaccharide content of the artificial diet. The

Table 5.2 *Corcyra cephalonica*, *Spodoptera litura*, artificial diet, artificial diets + *Corcyra cephalonica*, and *Corcyra cephalonica* with water on the foregut and hindgut quantitative enzyme profiles of *Rhynocoris marginatus*

Treatments	Amylase	Invertase	Lipase	Protease
Foregut				
Corcyra cephalonica	0.24 ± 0.03	0.10 ± 0.02	1.1 ± 0.2	13.4 ± 1.6
Corcyra cephalonica + water	0.23 ± 2.5	0.96 ± 0.2	0.8 ± 0.0	10.2 ± 0.5
Artificial diet alone	0.25 ± 1.3	0.11 ± 1.3	1.2 ± 0.2	13.9 ± 0.3
Artificial diet + *Corcyra cephalonica*	0.23 ± 1.2	0.95 ± 0.2	0.9 ± 0.0	11.2 ± 0.2
Spodoptera litura	0.19 ± 2.2	0.12 ± 1.2	2.6 ± 1.2	12.0 ± 0.2
Hindgut				
Corcyra cephalonica	0.58 ± 2.5	0.67 ± 0.01	0.7 ± 0.2	5.2 ± 1.0
Corcyra cephalonica + water	0.48 ± 0.5	0.55 ± 0.5	0.8 ± 0.2	4.5 ± 0.5
Artificial diet alone	0.69 ± 0.7	0.72 ± 0.5	0.8 ± 0.6	6.9 ± 1.6
Artificial diet + *Corcyra cephalonica*	0.52 ± 1.2	0.55 ± 0.5	0.6 ± 0.4	4.5 ± 1.2
Spodoptera litura	0.44 ± 0.3	0.24 ± 1.2	1.5 ± 0.5	5.1 ± 0.3

Amylase activity based on mg glucose released/mg protein/min
Invertase activity based on μl maltose released/mg protein/min
Lipase activity based on μl NaOH released/mg protein/min
Protease activity based on μl casein released/min

digestion of glycogen, obtained from meat and arthropod prey, might also explain the presence of amylase in the gut. In vertebrates like pig, carbohydrate is stored as glycogen in the liver.

To digest the vertebrate glycogen, *Rhynocoris marginatus* secrete more amount of amylase than to digest insect-based carbohydrate. *Rhynocoris marginatus* ingest intact starch granules, with starch digestion occurring completely in the midgut. Also, the ingestion of soluble glycogen from prey may be an alternative or additional explanation for the presence of amylase in the gut and its absence in the salivary glands.

The distribution of amylase is either or both the salivary glands and guts of heteropterans in various ecological niches which remain to be clarified by Boyd et al. (2002). Controversially, plant-feeding mirids usually have high levels of amylase in the salivary glands Agusti and Cohen (2000). These amylase secretions are thought to be ingested by the mirid along with partially digested starches to be used in the midgut to continue the breakdown of starch (Hori 1973; Takanona and Hori 1974; Wheeler 2001). The lack of detectable amylase in the salivary glands of *Rhynocoris marginatus* indicates that plant material is not a part of diet; this fundamental enzyme is found in the salivary glands of phytophagous heteropterans (Hori 2000).

Predacious mirids, *Deraeocoris pulchellus* (Reuter) and *Deraeocoris punctulatus* (Fallen), have very low amylase activity in salivary glands (Hori 1972), because they ingest both plant sap and arthropods as their food. But *Deraeocoris nigritulus* predator has no detectable amylase activity in salivary glands. A geocorid, *Geocoris punctipes* (Say), showed amylase activity in salivary glands, indicating its ability to digest starches of plants before ingestion (Zeng and Cohen 2000a). Similarly *Orius insidiosus* (Say), a predacious anthocorid has amylase activity detected by whole-organism homogenization, demonstrating ability to digest starch (Zeng and Cohen 2000b).

Cohen (1996) showed that the predator, *Nabis alternatus* Parshley (Nabidae) and *Sinea confuse* Caudell, recorded lack of amylase activity in their salivary glands. In another reduviid, *Zelus renardii* shows amylase activity in salivary glands. Moreover potent salivary secretions amplify predaceous hemipterans food selectivity and their efficient utilization of prey materials. The presence of enzyme is abundant in three diets which implies that this enzyme has some important physiological role in insect digestion of food materials (Boyd 2001, 2003; Boyd et al. 2002; Azevedo et al. 2007). There are low levels of amylase in the salivary gland and high levels in the hindgut of AC category, indicating that *Rhynocoris marginatus* might ingest disaccharides and digest them in the hindgut as proposed by Boyd (2003) in *D. nigritulus.*

5.4.2 Invertase

Sunderland (1988) method was adapted to estimate the invertase activity using 0.2 % sucrose as substrate in the reaction mixture, consisting of 10 Mm phosphate buffer of pH 6.8 and measuring the glucose at 30 °C. Moreover the glucose estimation was done using the dinitrosalicylic acid reagent with dextrose for this standard at 540 nm.

This study showed that irrespective of the categories, invertase activity was higher in the hindgut followed by the salivary gland and foregut. Artificial diet-reared predator hindgut showed maximum invertase activity (0.72 mg protein released/min), followed by *Corcyra cephalonica* (0.67 mg protein released/min). The invertase activity was increased from 0.6 mg protein released/min to 3.0 mg protein released/min in salivary glands when *Rhynocoris marginatus* take artificial diet. In foregut *Corcyra cephalonica* showed the lowest invertase activity (0.10 mg protein released/min) ($t = 1.176$, df = 2, $P = 0.360$) but increase the salivary gland (0.2 mg protein released/min) ($t = 40.06$, df = 2, $P = 0.001$). Among all the preys tested, *Spodoptera litura* reared reduviid salivary gland has the minimum activity (0.68 mg protein released/min).

Invertase appears to be an important enzyme for both plants and animals (Heil et al. 2004). Invertase commonly found in the secretion of saliva and digestive tract of various insects has been examined by many authors and was comprehensively reviewed by House (1965). Investigations on alimentary canal carbohydrate of nymphal predators have demonstrated that invertase was concentrated mainly in the foregut. Small quantity of invertase has been identified in the salivary gland. Twelve to 18 % of carbohydrate has been found in foreguts (Yazlovetsky 1992). The main activity volume of invertase was found to concentrate in foreguts. This data also confirm the functional role of *Rhynocoris marginatus* foreguts as the main zone for food digestion. Invertase also termed as beta-fructosidase, saccharadase, or sucrase catalyzes the cleavage of sucrose into the two monosaccharides, glucose and fructose (Naunoff 2001).

When *Spodoptera litura* was provided as prey, the salivary gland secretes more amount of invertase. It doubled and tripled the amount compared to artificial diet and *Corcyra cephalonica*, respectively. Since very less quantity of invertase was secreted in OC and CW at the salivary gland, the foregut showed maximum activity (0.95/0.96). The presence of invertase in the salivary glands suggests that *D. nebulosus* can break down the sugary water from plants before ingestion, unless this enzyme is a non-secretable cellular enzyme of the salivary glands as suggested by some heteropteran studies (Hori 2000). Twelve other mirids were found to have low levels of glucosidase in the salivary glands (Agusti and Cohen 2000; Hori 2000). Because the substrates of glucosidase are water soluble and commonly found might explain why only low levels of this enzyme are found in the salivary glands of many heteropterans. One notable exception is the coreids, which have strong glucosidase activity (Taylor and Miles 1994; Hori 2000; Boyd et al. 2002).

5.4.3 Lipase

Lipase is another digestive enzymes present in various parts of the insect intestine. Though enormous reports were available in phytophagous insects, scanty information is reported for predatory insects. Increasing experimental evidence indicates synthesis and control mechanisms for lipase complex, and there may be separate control over secretion and synthesis in predatory insects like reduviids. Lipase enzyme activity was estimated using olive oil emulsion as a substrate in the reaction mixture with 10 mM phosphate buffer (pH 7.2) and measuring the pH drops by about 0.2 units. 0.1N NaOH was added to bring the pH back to the original level. The amount of NaOH added was measured, and it is equivalent to the fatty acids (Jayaraman 1985) formed during the reaction.

Lipase activity was low in salivary gland of all the categories. In general *Spodoptera litura* fed reduviid foregut showed the maximum amount of lipase (2.6 mg/min) ($t = 61.36$, df-2, $P = 0.000$) than those fed with artificial diet (1.24 mg/min) ($t = 13.62$, df-2, $P = 0.005$) and *Corcyra cephalonica* (1.1 mg/min) ($t = 19.82$, df-2, $P = 0.003$) (Table 5.2). The salivary gland showed meager amount of lipase activity in all the categories. As observed in fore- and hindguts, *Spodoptera litura* (0.2) ($t = 48.32$, df-2, $P = 0.0005$) has the higher amount of lipase than reduviids reared on artificial diet (0.1 mg/min) ($t = 10.78$, df-2, $P = 0.008$) and *Corcyra cephalonica* (2.6 mg/min) ($t = 3.50$, df-2, $P = 0.073$).

Proteins are digested externally, but lipid is not hydrolyzed until the ingested material reaches the midgut where lipase activity rises after feeding the prey. The decrease in enzyme activity could be related with some anatomical and physiological modification of predator guts (Edwards 1961). *Rhynocoris marginatus* foregut has the maximum activity than other two parts of the insect tested. The prey paralysis may result from damage by certain digestive enzymes in neurons, muscles, and storage tissues of the prey. Structural proteins are hydrolyzed and liquefied inside the prey by endopeptidases such as trypsin-like injected to the predator, while cell membranes, storage tissues, and reproductive systems are affected by phospholipase, triacylglycerol lipases, and α-amylase. Latter, the hydrolyzed material is ingested with other nutrients from the prey, and additional processing occurs in the gut allowing its absorption by the predator (Cohen 1993, 1995). So in heteropteran, lipase enzymes are among the key enzymes for digestion and also predation.

Predacious arthropods lack extraintestinal digestive glands, the source of the entire digestive enzymes is the gut, and therefore the extra-oral digestion mechanism is a refluxing type (Cohen 1993, 1998a). Amylase, protease, lipase, and trypsin activities were detected in the entire salivary gland of *Rhynocoris marginatus*. Amylase, invertase, lipase, and protease activities ranged from 0.12 to 0.17, 0.15 to 0.6, 0.1 to 0.3, and 1.4 to 2.5, respectively. Enzymes of the salivary glands showed that *Rhynocoris marginatus* could be characterized as a zoophagous animal. *Corcyra cephalonica* was reared with starch rice cereals; *Rhynocoris marginatus* use amylase of their salivary glands to obtain energy from the starch of cereals ingested by this pest. It is also involved in the glycogen mobilization (Azevedo et al. 2007).

5.4.4 Protease

Proteases or endopeptidases are one of the most important digestive enzymes which have prominent crucial roles in converting proteins to oligo- and dipeptides. In general these enzymes are classified based on amino acids in their active site and the site of activity on protein molecules. Proteinases are responsible for initial digestion of proteins by breaking internal bonds or peptide bonds. Because of the variance in peptide chains, different proteinases are necessary to break these bonds. These proteinases have been classified to three main subclasses according to their active site, namely, (1) serine, (2) cysteine, and (3) aspartic proteinases. In each of these subclasses, there are several proteinases differing in substrate specificities. The oligopeptides resulting from proteinase action are attacked from the N-terminal end by aminopeptidases and from the C-terminal end by carboxypeptidases; both enzymes liberate one amino acid residue at each catalytic step.

To quantify the level of total proteinase, a 60 μl of diluted enzyme was added to 200 μl of 1 % azocasein (in 0.2 M glycine–NaOH, pH 10.0) and incubated at 37 °C for 30 min. The reaction was terminated by the addition of 300 μl of 5 % trichloroacetic acid (TCA). After centrifugation at 15,000 g for 10 min, an equal volume of 1 M NaOH was added to the supernatant, and absorbance was measured at 450 nm. Determinations were always carried out in triplicate. Azocasein and tyrosine were used as substrate and standard as suggested by Brock et al. (1982).

The proteinases are a major group of hydrolytic enzymes in insects and are involved in digestive processes, proenzyme activation, liberation of physiologically active peptides, complement activation, and inflammation processes, among others. The proteinases are classified according to their mechanism of catalysis:

1. *Serine proteinases*
2. *Cysteine proteinases*
3. *Aspartic proteinases*
4. *Metalloproteinases*

Both in fore- and hindguts intense protease activities were observed in artificial diet (13.9 mg protein released/min) (t = 42.86, df-2, P = 0.001)-reared predator followed by *Corcyra cephalonica* (13.4 mg protein released/min) (t = 57.89, df-2, P = 0.000)-reared reduviids. Least protease activity was recorded in *S. litura* (12.0)-reared predator both in fore- and hindguts (5.1 mg protein released/min) (Table 5.2). A similar trend was also observed in the salivary gland of *Rhynocoris marginatus*, but *Corcyra cephalonica* has the maximum activity (3.0 mg protein released/min) followed by artificial diet (2.4 mg protein released/min) (t = 17.61, df-2, P = 0.003) and *Spodoptera litura* (0.4 mg protein released/min) (t = 15.02, df-2, P = 0.004).

Studies on distribution pattern for lipolytic enzymes yield conclusions similar to studies of invertase and amylase. The foregut and hindgut of *R. marginatus* have been found to lower amount than other enzymes. Ishaaya et al. (1971) while working on protease activities in the larvae of *Spodoptera litura* recorded that certain protein factors resent in food stimulate digestive enzymes. This decline may result from a greater degradation or lower synthesis of the digestive protease produced by a quantitative decrease of the food intake.

Protease activity of the foregut, hindgut, and salivary gland of *Rhynocoris marginatus* has been found to be higher than other enzymes and indicates that this reduviid is exclusively predacious. Among the four enzymes studied, protease content was high in the salivary gland. Therefore, the protease of the salivary gland of this predator may be involved in the amino acid mobilization of their prey. Maximum mobilization was took place in *Corcyra cephalonica.* The toxicity of the saliva of Heteroptera predators is due to digestive enzymes (Baptist 1941). The prey paralysis may result from damage by certain digestive enzymes in neurons, muscles, and storage tissues of the prey. The presence of protease activity in the alimentary canal and salivary gland complex of predator was expected. Maximum protease activity was detected in the foregut, but the hindgut and salivary gland had low protease activity.

The presence of protease enzymes demonstrates the insect's ability to access structural or other insoluble proteins (Cohen 1993, 1998a, 2000). The endoproteinase attacks protein residues. These enzymes are already identified and reported by Zeng and Cohen (2001) and Boyd et al. (2002) in the salivary gland of a reduviid, *Zelus renardii* Kolenati (Cohen 1998a). The strictly phytophagous Heteroptera *Poecilocapsus lineatus* (F.) lacks detectable digestive proteases in the gut and salivary gland complex (Cohen and Wheeler 1998) indicating its inability to use animal protein. However, the phytophagous mirids, *Lygus lineolaris* and *Lygus hesperus*, have trypsin-like protease and elastase-like enzymes but not chymotrypsin-like enzymes in their salivary gland complexes (Agusti and Cohen 2000), demonstrating their ability to use animal protein as their nutrients. *Lygus hesperus* produces more proteases when given an artificial diet (Zeng and Cohen 2001). Same observations were also observed both in fore- and hindguts intense protease activity in artificial diet (13.9 mg protein released/min) provided *Rhynocoris marginatus*. The green mirid, *Creontiades dilutus* (Stal), was the first mirid shown to have protease activity in the salivary glands (Colebatch et al. 2001).

The protease enzyme secreted on the food from the salivary gland through the salivary canal is thought to be injected to the predator along with food and employed in the gut to continue breakdown of proteins (Cohen 1998b). Low levels of activity of protease in the hindgut and salivary gland indicate either the ability of *Rhynocoris marginatus* to synthesize protease in the gut or the presence of symbiotic microorganism in the gut. These proteases help to break down proteins once they are inside the gut, making the proteins absorbable after further hydrolysis of exopeptidases such as aminopeptidase and carboxypeptidase (Boyd et al. 2002; Boyd 2003).

The presence of protease activity at the salivary gland complex of *Rhynocoris marginatus* indicates that this reduviid is primarily predacious. The presence of protease activity in alimentary canal and salivary gland enzymes demonstrates an insect's ability to access structural or other insoluble proteins (Cohen 1993, 1998a, 2000). Structural proteins are hydrolyzed and liquefied inside the prey by endopeptidases such as trypsin-like injected to the predator, while cell membranes, storage tissues, and reproductive systems are affected by phospholipases, triacylglycerol lipases, and α-amylase. Irrespective of the preys and meat-based oligidic diets, trypsin activity was moderate in the salivary gland of *Rhynocoris marginatus*.

Later, the hydrolyzed material is injected with other nutrients from the prey, and additional processing occurs in the gut allowing absorption by the predator (Cohen 1993, 1995). Since the insect has the capacity to alter midgut composition within the same generation for neutralizing the effect of diet, understanding of the nature of gut protease activity together with the activity induced upon feeding of diet will be important for selecting diet for insect growth development. This process of altering the secretion of proteases may be more complicated in generalized (polyphagous) feeders than the specialized feeders. The generalized feeders are shown to be more adapted to several classes of inhibitors as compared to the specialized feeders (Broadway and Villani 1995). Previous accounts of proteases from salivary glands of predaceous hemipterans are limited of a few reports on aquatic species (Rastogi 1962; Khan 1964; Rees and Offord 1969) and also in a venom-spitting reduviid, *Platymeris rhadamanthus* Gerstaecker (Edwards 1961). All species studied here showed definite predigest activity with regard to proteases and phospholipases and rather weak triacylglycerol lipases.

All the hunter reduviids are insect feeders, and this ability requires them to have a good complement of digestive enzymes so as to bring about hydrolysis of the complex nutrients that characterize insects. Reduviids inject toxic salivary secretion into the host during paralysis and consumed the body content of the victim (Sahayaraj 2004). Cohen (1998a, b) reported that predatory reduviids consume partially digested food. Before consuming the host, predators predigest the host with the help of the salivary enzyme and then ingest the partially digested proteins, carbohydrates, and lipids and other nutrients, which can be further digested with alimentary canal enzymes. The protease activity of the gut wall is relatively low and does not appear to fluctuate with feeding. Since only trypsin digestion seems essential for protein absorption, it is possible that the protein breakdown affected by the saliva during external digestion and subsequent storage in the capacious midgut crop of the assassin bug is sufficient to yield assimilable products.

The biochemical and physical environments of the digestive system of *Rhynocoris marginatus* may be changed appreciably by the presence, absence, or quantity of a compound or mixture of compounds. With ingestion, induced or introduced changes in the gut environment may result in changes in pH, reducing potential, conductivity of specific ions, etc., or in change in activities of many enzymes such as protease or mixed function oxidases.

5.5 Salivary Gland Enzyme Profile

Enzymes are common components of several venomous arthropods including bees, wasps, scorpions, and spiders. In this respect, the saliva of reduviid predators is no different; it contains a wide variety of enzymes (says Sahayaraj et al. 2013). However, previously, in 1961, Edward reported that salivary amylase, lipase, hyaluronidases, and phospholipase are found in Cimicomorpha (Reduviidae) and, as expected, were detected in saliva based on substrate-specific enzyme assays. The trypsin-like enzyme in the saliva permits extra-oral digestion for use of protein in animal food. Trypsin-like enzymes and amylase are very active in the salivary glands of the reduviid predator *Zelus renardii* (Cohen 1993). *Rhynocoris marginatus* saliva shows acid and alkaline phosphatase activity, which is consistent with venom from the parasitic wasp *Pimpla hypochondriaca* (Dani et al. 2005).

Results reveal that high level of invertase, protease, amylase, and lipase in salivary gland is responsible for preoral digestion of this predator (Table 5.3). Adults of *Rhynocoris marginatus* salivary gland contain high level of protease, lipase, invertase, and amylase when reared on *Corcyra cephalonica*, *Corcyra cephalonica* with water and artificial diets + *Corcyra cephalonica*, *Spodoptera litura*, and artificial diet separately.

The presence of amylase in the salivary gland complexes of *Rhynocoris marginatus* indicates a disposition of this species toward utilization of nutrients that can be derived only from plants, either directly from ingestion of plant macromolecules or from second-hand ingestion of plant material from the digestive system of their

Table 5.3 *Corcyra cephalonica*, *Spodoptera litura*, artificial diet, artificial diets + *Corcyra cephalonica*, and *Corcyra cephalonica* with water on the salivary gland quantitative enzyme profiles of *Rhynocoris marginatus*

Treatments	Amylase	Invertase	Lipase	Protease
Corcyra cephalonica	0.17 ± 0.01	0.20 ± 0.04	0.2 ± 0.05	3.0 ± 1.1
Corcyra cephalonica + water	0.13 ± 0.5	0.15 ± 0.5	0.3 ± 0.1	2.5 ± 0.4
Artificial diet alone	0.18 ± 0.4	0.30 ± 0.8	0.1 ± 0.1	2.4 ± 0.1
Artificial diet + *Corcyra cephalonica*	0.12 ± 0.2	0.15 ± 0.05	0.3 ± 0.5	2.5 ± 0.2
Spodoptera litura	0.14 ± 0.3	0.60 ± 0.1	0.2 ± 0.0	1.4 ± 0.2

Amylase activity based on mg glucose released/mg protein/min
Invertase activity based on μl maltose released/mg protein/min
Lipase activity based on μl NaOH released/mg protein/min
Protease activity based on μl casein released/min

prey. Protease and amylase level has not been changed while artificial diet was provided to the predator. Lipase level enhanced when *Corcyra cephalonica* was provided along with artificial diet and shows provision of artificial diet has not much impacted on the digestive physiology of the reduviids.

The artificial diet may have biochemical and chemical effects on metabolic systems of each developmental stages of the insect. The present work is an important attempt dealing with only the dietary implication on insect with microbial population. Therefore it would be reasonable to suggest that the effect of the diet on survival and population may be related to their effect on the nutritional value of the artificial diet and natural and factitious prey. These data mostly indicate that artificial diet affect microbial population may be of practical importance in artificial mass rearing of reduviids predator. If it decreases the antimicrobial agents and other chemical agents, a wider variety of antimicrobial agents and chemical shall be screened for use in the nymphal diet of the insect.

Predatory insects are known to be more sensitive than their host, which are free-living insects (Grenier 1977). If antimicrobial agents added into the artificial diet of *Rhynocoris marginatus* nymphs weaken the performance of adult individuals, this may become a more important problem in biological control programs. So although there were low levels of microbial population in artificial diet-reared insects, it did not affect the survival and development of the insect in natural and factitious prey. To investigate the physiological aspects of endosymbiosis development of a nutritionally defined artificial diet for *Rhynocoris marginatus* is needed.

5.6 Conclusions

The characterization of the abovementioned enzymes in salivary secretions as well as in alimentary canal secretion of reduviid could give a background of predatory association of the bug with prey. It also shows the adaptation of the reduviid against the artificial diet feeding. Enzyme secretion is location specific as well as food specific both at salivary gland and gut.

5.7 Future Recommendations

- Along with studied enzymes, it is necessary to study other enzymes of both the intestine and salivary gland.
- Basic differences in the mechanism of synthesis, storage, and release of the various digestive enzymes at the cellular level are needed to be described in various reduviids in relation to natural hosts, laboratory insect species, and also artificial diets.
- It is essential to study the details of gut international structures by electron microscope.
- Levels of different enzymes in relation to optimum temperature and pH and general and specific inhibitors could be studied with high technologies.

References

Agusti N, Cohen AC (2000) *Lygus hesperus* and *L.lineolaris* (Hemiptera: Miridae), phytophages, zoophages, or omnivores: evidence of feeding adaptations suggested by the salivary and midgut digestive enzymes. J Entomol Sci 35:176–186

Alomar O, Wiedenmann RN (1996) Zoophytophagous Heteroptera: implications for life history and integrated pest management. Thomas Say Publications in Entomology. Entomological Society of America, Lanham

Ambrose DP, Maran SPM (2000) Polymorphic diversity in salivary and haemolymph proteins and digestive physiology of assassin bug *Rhynocoris marginatus* (Fab.) (Heteroptera : Reduviidae). J Appl Entomol 124:315–317. doi:10.1046/j.1439–0418.2000.00473.x

Azevedo DO, Zanuncio JC, Zanuncio JS, Martins GF, Marques-Silva S, Sossai MF, Serrao JE (2007) Biochemical and morphological aspects of salivary glands of the predator *Brontocoris tabidus* (Heteroptera: Pentatomidae). Braz Arch Biol Technol 50(3):469–477

Baker JE (1991) Purification and partial characterization of amylase allozmes from lesser grain borer, *Rhyzopertha dominica*. Insect Biochem 21:303–311

Balasubramanian TP, Lakshmana Perumal Swamy D, Mohan C, Natarajan R (1975) Cellulolytic activity of marine streptomycetes isolated from the alimentary canal of marine borer. 16th annual conference of AMI, held at virbadhra (U.P.)

Balogun RA, Fisher O (1970) Studies on the digestive enzymes of the common african toad, *Bufo regulasis* bonlenger. Comp Biochem Physiol 33:813–820

Baptist BA (1941) The morphology and physiology of the salivary glands of Hemiptera-Heteroptera. Q J Microsc Sci 83:91–139

Boyd DW Jr (2003) Digestive enzymes and stylet morphology of *Deraeocoris nigritulus* (Uhler) (Hemiptera: Miridae) reflect adaptations for predatory habits. Ann Entomol Soc Am 96(5):667–671

Boyd DW, Jr (2001) *Deraeocoris nebulosus* (Uhler) (Hemiptera: Miridae): a potential biological control agent. Ph.D. dissertation, Clemson University, Clemson, SC

Boyd DW Jr, Cohen AC, Alverson DR (2002) Digestive enzymes and stylet morphology of *Deraeocoris nigritulus* (Uhler) (Hemiptera: Miridae) a predacious plant bug. Ann Entomol Soc Am 95:395–401

Broadway RM, Villani MG (1995) Does host range influence susceptibility of herbivorous insects to non-host plant proteinase inhibitors? Entomol Exp Appl 76(3):303–312

Brock RM, Forsberg CW, Buchanan-Smith JG (1982) Proteolytic activity of rumen microorganisms and effect of proteinase inhibitors. Appl Environ Microbiol 44:561–569

Cobben RH (1979) On the original feeding habits of the Hemiptera (Insecta): a reply to Merrill Sweet. Ann Entomol Soc Am 72:711–715

Cohen AC (1990) Feeding adaptations of some predaceous Hemiptera. Ann Entomol Soc Am 83(6):1215–1223

Cohen AC (1993) Organization of digestion and preliminary characterization of salivary trypsin-like enzymes in a predaceousheteropteran, *Zelus renardii*. J Insect Physiol 39:823–829

Cohen AC (1995) Extra-oral digestion in predaceous terrestrial Arthropoda. Annu Rev Entomol 40:85–103

Cohen AC (1996) Plant feeding by predatory Heteroptera: evolutionary and adaptational aspects of trophic switching. In: Alomar O, Wiedenmann RN (eds) Zoophytophagous heteroptera: implicationsfor life history and integrated pest management. Thomas Say Publicationsion Entomology. Entomological Society of America, Lanham, pp 1–17

Cohen AC (1998a) Solid-to-liquid feeding: the inside(s) story of extra-oral digestion in predaceous Arthropoda. Am Entomol 44:103–117

Cohen AC (1998b) Biochemical and morphological dynamicsand predatory feeding habitsin terrestrial heteroptera. In: Ruberson JR, Coll M (eds) Predaceous heteroptera: implications for biological control. Thomas Say Publications in Entomology. Entomological Society of America, Lanham, pp 21–32

Cohen AC (2000) Feeding fitness and quality of domesticated and feral predators: effects of long-term rearing on artificial diet. Biol Control 13:49–54

Cohen AC, Wheeler AG Jr (1998) Role of saliva in the highly destructive fourlined plant bug (Hemiptera: Miridae: Mirinae). Ann Entomol Soc Am 91:94–100

Colebatch G, East P, Cooper P (2001) Preliminary characterisation of digestive proteases of the green mirid, *Creontiades dilutus* (Hemiptera: Miridae). Insect Biochem Mol Biol 31:415–423

Coolbear T, Danial R, Morgan (1992) The enzymes from extreme thermophilies, bacteria sources, thermo strains identified as extremely thermophilic bacilli for extra cellular proteolytic activity and general property of proteinases from two of the strains. J Appl Bacterialogy 71:525

Dani MP, Edwards JP, Richards EH (2005) Hydrolase activity in the venom of the pupal endoparasitic wasp, Pimpla hypochondriaca. Comp Biochem Physiol B Biochem Mol Biol 141(3):373–381

Edwards JS (1961) The action and composition of the saliva of an assassin bug *Platymeris rhadamanthus* Gaerst. (Hemiptera: Reduviidae). J Exp Biol 8:61–77

Grenier S (1977) Effets nocifs de la nipagine M sur le parasitoïde Phryxe caudata [Dipt.: Tachinidae]. Entomophaga 22:23–26

Heil M, Rita buchler R, Boland W (2004) Quantification of invertase activity in ants under field conditions. J Chem Ecol:RC177–RC183

Hori K (1969) Effect of various activators on the salivary amylase of the bug *Lygus disponsi*. J Physiol 15:2305–2317

Hori K (1972) Comparative study of a property of salivary amylase among various heteropterous insects. Comp Biochem Physiol 42B:501–508

Hori K (1973) Studies on enzymes, especially amylases, in the digestive system of the bug *Lygus disponsi*and starch digestion in the system. Res Bull Obihiro Univ 8:173–260

Hori K (2000) Possible causes of disease symptoms resulting from the feeding of phytophagousHeteroptera. In: Schaefer CW, Panizzi AR (eds) Heteroptera of economic importance. CRC, Boca Raton, pp 11–35

House HL (1965) Digestion. In: Rockstein M (ed) Physiology of insecta, vol 2. Academic, New York, pp 818–858

House HL (1967) Artificial diets for insects. A compilation of references and abstracts, Information Bulletin No. 5. Research Institute Canada Dept. Agric, Ontario

Ishaaya I, Moore I, Joseph B (1971) Protease and amylase activity in the larvae of the Egyprioan cotton worm, *Spodoptera littoralis*. J Insect Physiol 17:945–953

Jayaraman J (1985) Laboratory manual in biochemistry. Wiley Eastern Ltd., New Delhi. pp 75–76, 107

Khan MA (1964) Proteolytic activity in the digestive tract of the water scorpion *Laccotrephis maculatus* Fab. (Nepidae: Hemiptera). Entomol Experimetalis Applic 7:335–338

Miles PW (1972) The saliva of hemiptera. Adv Insect Physiol 9:183–256

Naunoff DG (2001) β-Fructosidase superfamily: homology with some α-L-arabinases and β-D-xylosidases. Proteins: Struct Func Bioinf 42(1):66–76

Nigam CS, Omkar (2003) Experimental animal physiology and biochemistry. New Age International (p) Limited, Publishers, New Delhi

Rastogi SC (1962) On the salivary enzymes of some phytophagous and predaceous heteropterans. Sci Cult 28:479–480

Ravikumar T, Albert S, Sanjayan PK (2002) Efficiency of digestion of oligosaccharides in the gut and salivary gland of some seed feeding lygaeids (Heteroptera : Lygaeidae). Entomon 27(1):43–50

Rees AR, Offord RE (1969) Studies on the protease and other enzymes from the venom of *Lethocerus cordofanus*. Nature 221:665–667

Sahayaraj K (2004) Reduviids in biological control. In: Sahayaraj K (ed) Indian insect predators in biological control. Dayas Publication, India, pp 134–166. ISBN 8170353408

Sahayaraj K (2007) Isolation, identification and characterization of gut flora of three reduviid predators. Asian J Microbiol Biotechnol Environ Sci 9(4):1073–1075

Sahayaraj K (2014) Reduviids and their merits in biological control. In Basic and applied aspects of biopesticides. Springer, India, p 195–214

Sahayaraj K, Balasubramanian R (2008) Biological control potential evaluation of artificial and factitious diets reared *Rhynocoris marginatus* (Fab.) on three pest. Arch Phytopathol Plant Protect 42(3):238–247

Sahayaraj K, Venkatesh P, Balasubramanian R (2007a) Feeding behaviour and biology of a reduviid predator *Rhynocoris marginatus* (Fabricius) (Heteroptera: Reduviidae) on Oligidic Diet. Hexapoda 14(1):24–30

Sahayaraj K, Kumara Sankaralinkam S, Balasubramanian R (2007b) Prey influence on the salivary gland and gut enzymes qualitative profile of *Rhynocoris marginatus* (Fab.) and *Catamiarus brevipennis* (Serville) (Heteroptera: Reduviidae). J Insect Sci 4(4):331–336

Sahayaraj K, Vinoth Kanna A, Muthukumar S (2010) Gross morphology of feeding canal, salivary apparatus and digestive enzymes of salivary gland of *Catamirus brevipennis* (Servile) (Hemiptera: Reduvidae). J Entomol Res Soc 12(2):37–50

Sahayaraj K, River D, Muthukumar S (2013) Biochemical and electrophoretic analyses of saliva from the predatory reduviid *Rhynocoris marginatus* (Fab.). Acta Biochim Pol 60(1):91–97

Schaefer CW (1997) The origin of secondary zoophagy from herbivory in Heteroptera (Hemiptera). In: Raman A (ed) Ecology and evolution of plantfeeding insects in natural and man-made environments. International ScientiÞc Publications, New Delhi, pp 229–239

Simpson SJ, Raubenheimer D (1993) A multi-level analysis of feeding behaviour: the geometry of nutritional decisions. Philos Trans R Soc Lond B Biol Sci 342(1302):381–402.bb

Sunderland KD (1988) Quantitative methods for detecting invertebrate predation occurring in the field. Ann Appl Biol 112:201–224

Sweet MH (1979) On the original feeding habitsof hemiptera: Insecta). Annu Entomol Soc Am 72:575–579

Takanona T, Hori K (1974) Digestive enzymes in the salivary gland and midgut of the bug *Stenotus binotatus*. Comp Biochem Physiol 47A:521–528

Taylor GS, Miles PW (1994) Composition and variability of the saliva of coreids in relation to phytoxicoses 400 and other aspects of the salivary physiology of phytophagous heteroptera. Entomol Exp Appl 73:265–277

Tonapi TG (1996) Experimental entomology. CBS Publishers of Distributors, New Delhi

Wheeler AG Jr (2001) Biology of the plant bugs (Hemiptera: Miridae): pests, predators, opportunists. Cornell University Press, Ithaca

Wigglesworth VB (1972) The principles of insect physiology, 7th edn. Chapman and Hall, London, p 827

Xie ZN, Nettles WC Jr, Morrison RK, Irie K, Vinson SB (1986) Three methods for the in vitro culture of *Trichogramma pretiosum* Riley. J Entomol Sci 21:133–138

Yazgan S, House HL (1970) An hymenopterous insect, the parasitoid *Itoplectis conquisitor* reared axenically on a chemically defined synthetic diet. Can Entomol 102:1304–1306

Yazlovetsky IG (1992) Development of artificial diets for entomophagous insects by under standing their nutrition and digestion. In: Anderson TE, Leppla NC (eds) Advances in insect rearing for research and pest management. Westview Press, Boulder, pp 41–62

Zeng F, Cohen AC (2000a) Comparisonof amylase and protease activities of a zoophytophagous and two phytozoophagous Heteroptera. Comp Biochem Physiol 126A:101–106

Zeng F, Cohen AC (2000b) Demonstration of amylase from the zoophytophagous anthocorid *Orius insidiosus*. Arch Insect Biochem Physiol 44:136–139

Zeng F, Cohen AC (2001) Induction of elastase in a zoophytophagous heteropteran, *Lygus Hesperus* (Hemiptera: Miridae). Ann Entomol Soc Am 94:146–151

Body Total Protein and Genomic DNA

6

Abstract

Predatory reduviids consume partially digested food. Before consuming the host, predators predigested the host with the help of the salivary enzyme. In order to understand interaction pathways within complex food webs, it is necessary to characterize encounter frequencies between each constituent part of the web and describe the consequence of these interactions. The overall mean protein concentration was high in *Spodoptera litura* followed by *Rhynocoris marginatus*, *Dysdercus cingulatus*, and pig liver and *Corcyra cephalonica*. Six primers were tested to amplify the DNA of the predator. A total of 52 amplicons were obtained in which four primers in five populations of *Rhynocoris marginatus* were analyzed. The electrophoretic pattern reveals polymorphism occurred among predator populations. DNA polymorphism with 450 bp was common among three diets such as *Corcyra cephalonica* weekly once with water, oligidic diet and weekly once *Corcyra cephalonica*, and *Spodoptera litura*. Among the six primers, KTG-3 and KTG-5 have amplified better than other primers. The level of genetic diversity is far greater in the oligidic diet-reared *Rhynocoris marginatus*. *Rhynocoris marginatus* fed with *Spodoptera litura* and OD-reared predator yielded consistently high absorbance values followed by *Corcyra cephalonica* fed

Keywords

Body protein profiling • SDS-PAGE • Polypeptides • Genomic DNA profiling • Genetic analysis • RAPD analysis of reduviid predator

K. Sahayaraj, R. Balasubramanian, *Artificial Rearing of Reduviid Predators for Pest Management*, DOI 10.1007/978-981-10-2522-8_6

6.1 Introduction

All the hunter reduviids are insect feeders, and this ability requires them to have a good complement of digestive enzymes so as to bring about hydrolysis of the complex nutrients that characterize insects. Reduviids inject toxic salivary secretion into the host during paralysis and consume the body content of the victim (Sahayaraj 2004). Cohen (1998b) reported that predatory reduviids consume partially digested food. Before consuming the host, predators predigest the host with the help of the salivary enzyme.

In order to understand interaction pathways within complex food webs, it is necessary to characterize encounter frequencies between each constituent part of the web and describe the consequence of these interactions. Determination of encounter frequencies is, in general, relatively straightforward. Sampling protocols are routinely designed for monitoring both vertebrate and invertebrate populations, but measuring trophic connections between species can be difficult, especially in highly mobile and/or cryptic organisms. Then ingest the partially digested proteins, carbohydrates and lipids, and other nutrients, which can be further digested with alimentary canal enzymes. The protease activity of the gut wall is relatively low and does not appear to fluctuate with feeding. Since only trypsin digestion seems essential for protein absorption, it is possible that the protein breakdown affected by the saliva during external digestion and subsequent storage in the capacious midgut crop of the assassin bug is sufficient to yield assimilable products.

The biochemical and physical environments of the digestive system of *Rhynocoris marginatus* may be changed appreciably by the presence, absence, or quantity of a compound or mixture of compounds. With ingestion, induced or introduced changes in the gut environment may result in changes in pH, reducing potential, conductivity of specific ions, etc., or in change in activities of many enzymes such as protease or mixed functions of oxidases.

6.2 Total Macromolecule Content

All consumed foods are digested with anabolic, catabolic, and metabolic processes; fundamental and functional units are absorbed and utilized for growth, development, and reproduction via various metabolic pathways. Macromolecular metabolism, carbohydrates (cellobiose, fructose, galactose, glucose, lactose, maltose, melibiose, rassinose, sucrose, and trehalose), proteins, and lipids are generally considered as macromolecule constituents of both vertebrate and invertebrate including insects.

Totally ten uniformly sized preys *Corcyra cephalonica* and *Spodoptera litura* and 5 g pig liver were used for the macromolecule profile analyses. Total carbohydrates were estimated using glucose as a substrate (Nigam and Omkar 2003). Total protein was determined using the Lowry's method. Gut particles were removed

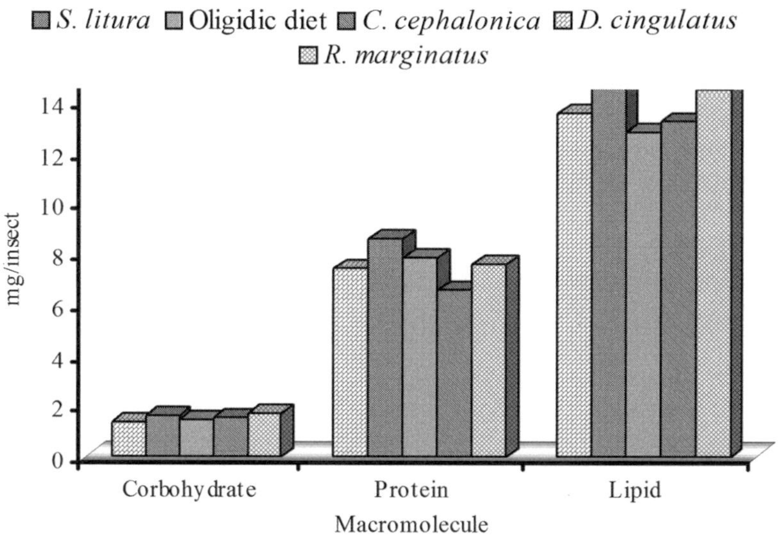

Fig. 6.1 Quantitative macromolecule components of pest, predator, and oligidic diet (in mg/100 mg)

from aqueous homogenates by centrifugation for 10 min at 15,000 rpm. Protein was quantified from a standard curve by using bovine serum albumin (BSA) as a standard at 580 nm. Lipid was estimated using olive oil as a standard (Jeyaraman 1981).

Whole-body total macromolecular content of the tested preys and oligidic diet is presented in Fig. 6.1. From the results, it was very clear that total carbohydrate, protein, and lipid contents were higher in *Spodoptera litura* followed artificial diet and *Corcyra cephalonica*.

Carbohydrates constitute an integral part of the exoskeleton of insects, and their role in physiological events such as molting, metamorphosis, flight, and diaphase has been well documented. Both glycogen and trehalose are important carbohydrates of insects, and hence many researchers considered them for their studies. The concentration of carbohydrates and other biochemical parameters mainly depends on the quality of plants and animals in phytophagous and zoophagous insects, respectively. In addition to carbohydrate, lipids are of vital importance to many insects as substrates for embryogenesis, metamorphosis, and flight.

Some nutrients obtained from feeding are stored in the fat body of the insects (an organ analogous with mammalian liver), which is a multifunctional organ and considered to participate in the metabolic processes with high biosynthetic activity. Further, it is the place of intense biosynthetic activities. Throughout the insect life, it is the main source for the hemolymph protein. Hemolymph proteins play an important role in insects for transport functions as well as their enzyme action. Proteins are an integral part of the cuticle; ant hemolymph plays an important role in metamorphosis and insect growth.

6.3 Protein Profiling

Rhynocoris marginatus, is a generalist predator of insects and has been used in biological control. However, information on the macromolecular profile as well as the whole-body protein profile in this insect is lacking. Therefore, we have studied the whole-body protein profile in *Rhynocoris marginatus* and further characterized the same.

Each insect was homogenized in ice-cold phosphate-buffered saline (PBS 500 μl) for 1 min with a homogenizer (Remi homogenizer, Mumbai) in 0.9 ml of phosphate-buffered saline containing 0.1 ml of protease inhibitors (Genei, Bangalore). The extracted aliquots were stored at a −20 °C (Inlab Equipments Pvt. Ltd., Chennai) until assayed. Total protein was determined using the Lowry's method discussed earlier. Gut particles were removed from aqueous homogenates by centrifugation for 10 min at 15,000 rpm. Protein was quantified from a standard curve by using bovine serum albumin as a standard. Color was read by a spectrophotometer at 580 nm and results were discussed. Individuals of *Rhynocoris marginatus* were homogenized in 500 μl of PBS. An aliquot from each sample from the first feeding trial was assayed by the immunoassays by the indirect ELISA described below. The optimum sample (500 μl) volume used for immunoassays was *Spodoptera litura*, *Dysdercus cingulatus*, and *Corcyra cephalonica* and oligidic diet.

6.3.1 SDS-PAGE for Protein Profiling

The protein profiles were studied by SDS-PAGE according to the method of Laemmli (1970). A 7.5 % slab gel was cast between the glass plates supplied by Genei, Bangalore. A 5 % stacking gel (pH 6.8) was also cast over the resolving gel. The protein samples were mixed with an equal volume of sample buffer (0.125 M Tris-HCl, pH 6.8 %, 4.6 % SDS, 20 % sucrose, 10 % β-mercaptoethanol, and 0.1 % bromophenol blue) and boiled for 2–3 min and then centrifuged at 10,000 g for 5 min, and 25 μl each of marker and samples were loaded onto the well. In each lane the samples containing 100–125 μg protein concentration were loaded. Initially, a current was applied at a constant voltage of 50 V till the samples reached the resolving gel and then increased to 100 V till the completion of run using electrophoresis power pack. Standard protein molecular weight markers (low molecular weight, Gene, Bangalore) were also run in one lane parallel to the samples. The gels were fixed in Coomassie Brilliant Blue (CBB) (ethanol/acetic acid/water 40:50:10) staining solution overnight and destained using ethanol/acetic acid/water (40:50:10) till the bands were clearly visible.

The appearance of an increased number of polypeptides and change in their mobility as well as intensity after OD for rearing indicated that ingredients of the AD stimulate and also initiate the synthesis of some new proteins in *Rhynocoris marginatus*. These findings get corroborated by the results of Mjeni and Morrison

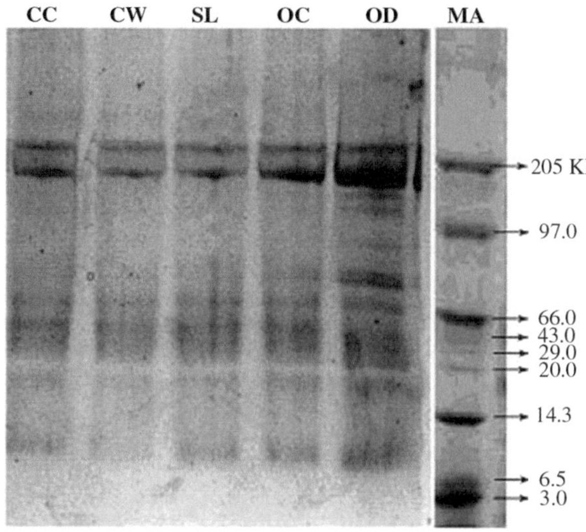

Plate 6.1 Whole-body protein profile and SDS profile of *Rhynocoris marginatus* (oligidic diet, *Corcyra cephalonica*, *Corcyra cephalonica* with water, *Spodoptera litura*, oligidic diet + *Corcyra cephalonica*, oligidic diet, and marker protein-MA)

(1976), Kajiura and Yamashita (1989), and Bradfield et al. (1990) as they also recorded an increase in the number of protein bands in the AD-reared adults (Plate 6.1).

SDS-PAGE separation of the gut and whole-body homogenate of the *Rhynocoris marginatus* in a mini gel (10 × 7.5 cm and 5 % and 10 % acrylamide for stacking and resolving gels, respectively). The electrophoretic separation of the denatured proteins from the whole gut and whole-body homogenate of *Rhynocoris marginatus* is shown in Plate 6.4. The PAGE analysis of proteins of predator revealed that AD influenced the number of bands and intensity of bands. A greater number of protein bands are visible in the separation of the whole-body homogenate proteins from insect fed on meat-based diet lanes 5 and 6 versus when insect was fed with control diet *Corcyra cephalonica*, *Corcyra cephalonica* weekly once with water lanes (1, 2), and natural prey *Spodoptera litura* lane 3. There are three distinct molecular weights of proteins ranged from 205 to 97 kDa in AD.

A dendrogram generated by hierarchical clustering techniques. Clustering analysis is a standard method to study relatedness and genetic diversity among the different diet-reared *Rhynocoris marginatus*. Five diet-reared *Rhynocoris marginatus* were resolved into five clusters based on the protein expression. *Corcyra cephalonica* and *Spodoptera litura*-reared reduviids cluster under one group, and their total similarity index was 77.5 %. Oligidic diet-reared reduviids and *Corcyra cephalonica* + weekly once with water categories protein cluster under the one group. The similarity index was 55 %. OD-reared category clusters with the above said category with 66 % indices. It is clear from the dendrogram that the reduviids reared in *C. cephalonica*, *Spodoptera litura*, and oligidic diet are the most divergent in their ability to store amino acids as energy source.

6.4 Genomic DNA Profiling

Rapid PCR-based screening systems for the study of the prey diversity of generalist predators have been developed to expand the potential of molecular detection into various areas of food web research. The techniques were described by Harper et al. (2005) using a PCR to simultaneously amplify DNA from a range of prey species. The retention time for DNA within the gut of a predator during digestion is influenced by factors including the size of the target DNA molecule. Predation by the coccinellid beetle *Coleomegilla maculata* De Geer (Coleoptera: Coccinellidae) upon the eggs of the European corn borer *Ostrinia nubilalis* (Hubner) (Lepidoptera: Crambidae) has been characterized by PCR amplification of four fragments of prey genomic DNA of different sizes from predator guts (Hoogendoorn and Heimpel 2001, 2002). The use of PCR in the gut content analysis of predator (Agusti et al. 1999, 2000; Zaidi et al. 1999; Chen et al. 2000; Hoogendoorn and Heimpel 2001) was studied by many workers.

Amplifying prey DNA from predator gut contents is used increasingly to elucidate predator–prey relationships and is comparable to serology in many aspects (Symondson 2002; Dodd et al. 2003). For instance PCR-amplified DNA of noctuid larvae by mirids (Agusti et al. 1999); aphid predation by a spider, ladybird, and lacewing larvae (Chen et al. 2000; Greenstone et al. 2005); European corn borer predation by a coccinellid (Hoogendorn and Heimpel 2001); and mosquito predation by a dragonfly (Morales et al. 2003) were available in the literature.

Identification of the gut contents of predatory insects can provide information on trophic relationships and the dynamics of predator–prey interactions. Several problems may be encountered in determining the diet of predatory insects in the field. Direct observations of predation (Legaspi et al. 1996; Munyaneza and Obrycki 1998) can be complicated by the fact that both prey and predator are often small and cryptic. Microscopic analysis of gut contents (Aussell and Linley 1994; Powell et al. 1996; Sleaford et al. 1996; Triltsch 1997) was possible for predators that ingest relatively large prey fragments. This work builds upon a number of recent studies on the use of PCR in the analysis of predator gut contents (Agusti et al. 1999; Agusti and Cohen 2000; Zaidi et al. 1999; Chen et al. 2000; Hoogendoorn and Heimpel 2001).

The pantropical Salyavatinae (Hemiptera: Reduviidae; 17 genera, 107 species) contains members with enigmatic morphology and specialized behavior for feeding on termites. All Salyavatinae are suspected specialist termite predators; however, existing observations are limited to seven species. Prior analyses indicate that Salyavatinae may be paraphyletic with respect to another subfamily, Sphaeridopinae, also hypothesized to feed on termites. A molecular phylogeny of these putative termite assassins is here constructed using seven loci from 28 species in nine genera and is used in a dating analysis to shed light on the timing of Neotropical colonization by this primarily Old World clade (Gordon and Weirauch 2015). Further, they reported that DNA extracted from gut contents of 50 individuals was assayed using

PCR with prey-specific primers. Molecular assays, along with recent photographs and observations, provide substantial evidence that this clade feeds specifically upon termites, documenting 28 new individual associations. Our phylogeny supports a sister group relationship of the Neotropical genus *Salyavata* with Sphaeridopinae. Termite association data combined with our phylogeny provide evidence of previously unknown prey conservatism among clades of one of the most diverse groups of specialist termite predators. Similarly, specific PCR for DNA of the pest, the kudzu bug *Megacopta cribraria* (F.) (Hemiptera: Plataspidae), of soybean by a reduviid *Zelus renardii* (Kolenati), along with nine other predators, was tested by Greenstone et al. (2014).

6.4.1 Genomic DNA Extraction

Reagents Required Sodium dodecyl sulfate (SDS), lysis buffer, ammonium acetate, isopropyl alcohol, 70 % ethanol, TE buffer.

Procedure For DNA extraction, insects were stored in 70 % ethanol, and DNA was extracted by drying the samples prior to extraction. Sambrook et al. (1989) method was followed for DNA extraction with some modifications as described below.

Rhynocoris marginatus was taken in a test tube and kept on ice. Add 500 μl of TEN (Tris/EDTA/NaCl) extraction buffer/2 % SDS (9:2) and 0.1 volume (50 μl) of proteinase K. Mix the solutions well and the total volume was made up to 550 μl. Using the homogenizer the sample was gently homogenized. Incubate the homogenate at 55–60 °C for 1 h. Add 500 μl of phenol (pH 8.00) and 448 μl of chloroform and 2 μl of isoamyl alcohol (25:24:1) (IAA) and shake for 10 min. Centrifuge the samples at 13,000 rpm for 10 min using a microcentrifuge. The lower (organic) phase was disposed of in suitable phenol waste bottles. 500 μl of chloroform and isoamyl alcohol (1:1) preparation was added to the sample and shaken gently for 10 min.

Remove the upper aqueous phase and place this into a new Eppendorf tube. Take 50 μl of 2 M sodium acetate (pH 5.6) and 1 ml of ice-cold ethanol (100 %) to the aqueous solution. The solution is inverted several times. Keep the samples at −20 °C for overnight incubation. Centrifuge the sample at 13,000 rpm for 15 min. The supernatant is discarded by blotting out with Whatman paper, and the precipitate is washed in 1 ml of 70 % ethanol. The washing step involved centrifuged at 13,000 rpm for 5 min with the pellet of DNA being on the inner surface during the spin. The alcohol was tipped out after each spin and blotted on a clean tissue, and then the remaining supernatant was removed using a fine pipette attached to a vacuum suction. Thereafter, the DNA pellet was air-dried for 15–30 min in room temperature. When the tubes were seen to be dry, the pellet was resuspended in 20 μl TE buffer and stored at −20 °C up to 1 month.

6.4.2 Quantification of DNA

The quantification of DNA was performed in UV–visible spectrophotometer (Eppendorf, Bangalore), in order to find out the purity as well as the quantity of the DNA.

Principle Conjugated double bonds in nitrogenous bases of nucleic acids absorb UV light strongly at 260 nm. Based on the amount of UV light absorbed at 260 nm, the nucleic acid concentration was recorded.

The purity of DNA was calculated by using the formula:

$$\text{The } 260/280\,\text{nm absorbency ratio also indicates the purity of DNA preparation.}$$
$$\text{The } 260/280 \text{ ratio for pure DNA} = 1.8$$

The amount of DNA was calculated by using the formula:

$$\text{Amount of DNA} = \text{OD at } 260\text{nm} \times 50 \times \text{Df}/1000$$
$$1\text{OD} = 50\mu\text{g of DNA.}$$

6.5 Genetic Analysis

6.5.1 Polymerase Chain Reaction (PCR)

A 0.2 ml PCR tubes for each genomic DNA were labeled and tubes were arranged in the rack provided to hold the 0.2 ml tubes. Two microliters of template DNA (calf thymus) having the concentration of 50 ng was added to the tubes which is already labeled and kept in the PCR tube rack. The following reaction cocktail was prepared in 1.5 ml microcentrifuge tubes for a required number of reactions plus two reactions to compensate the pipetting loss. 23 μl of the reaction mixture was added to the 0.2 ml PCR tube, which was already loaded with 2 μl of template DNA making the final volume to 25 μl. The content was mixed, and the PCR reaction was performed in a thermocycler following the program given below.

6.5.1.1 Electrophoretic Analysis of Genomic DNA

Reagents Required Tris–boric acid–EDTA (TBE) buffer, gel-loading dye, ethidium bromide (ETBR)

Equipments Required Electrophoresis tank, power pack, voltage (50 V), gel documentation apparatus, UV transilluminator

6.5.1.2 Principle

Electrophoresis refers to the separation of macromolecules of different sizes by application of a constant electric field (50 V) onto the DNA fragments placed in a matrix of polymerized agarose. As the DNA molecule is negatively charged, and travels toward the anode, it is loaded at the cathode end. The speed of migration of the fragments has an inverse relation with the size of DNA. The separated fragments are visualized by staining the gels with an intercalating dye (ethidium bromide), which fluoresces under UV light. Acrylamide gels are used for separation of small fragments of DNA (5–500 bp). Agarose gels can resolve DNA fragments varying in size from 200 bp to about 50 kb depending upon the concentration of agarose in the gel.

For checking the amplified product, we run the product in 1.5 % agarose gel and stained with ethidium bromide. 1.5 % agarose gel was casted with suitable gel electrophoresis system along with ethidium bromide. 2 μl of 6× loading dye was added to 10 μl of the PCR product and mixed well. The gel was loaded with PCR product. The gel was run in 1× TBE at 50 V till the bromophenol blue dye front reach the end of the gel. The gel was viewed using UV transilluminator (Genei, Bangalore), and the gel was documented using photo-documentation system (Biotech, Yercaud, Tamil Nadu). Photograph document was preserved as hard copy, or the image was stored in the hard disk as a soft copy. The banding pattern was analyzed.

The extracted DNA from the experimental predators was subjected to PCR analysis using six primers. We used primer sets that amplify fragments of different lengths that each has a characteristic detection time, which made it possible to estimate the time since feeding from the number of bands that was detectable. A PCR strategy that targets diverse items and is transferable across study systems would be desirable in many situations. Such an approach is reliant on the use of more or less "universal" PCR primers, that is, PCR primers designed to anneal to target templates in as wide a range of taxa as possible. The use of universal primers on DNA extracted from samples with a dominant DNA template (i.e., diet samples) results in primarily the dominant template being amplified, which may not be of interest.

Employing universal primer analyses in these situations requires further manipulation of the DNA to exclude unwanted templates. Such approaches have been reported, but they may not be applicable outside of the system they were designed for without considerable further in silico and laboratory development. Consequently, like prey-specific approaches, the formulation of these universal methods was largely ad hoc, and they are not immediately widely applicable. Considering the importance, we selected KTG-3 (-5′-GTAGACCCGT-3′) and KTG-5 (5′-AACGCGCAAC-3′) (Genei scientific supplies, Bangalore) as primers for the study.

PCR reactions were performed in 25 μl of reaction mixtures containing 1 mM dNTP mix (5.0 μl), 1.0 μl template DNA (50 ng/μl), 10 mM RAPD primer (2.5 μl), 10× reaction buffer, 25 mM Mg Cl_2 (1.5 μl), 2.5 units Taq polymerase enzyme (5 U/μl), and sterile distilled water. Abovementioned 25 μl of reaction mixtures was placed in PCR tubes in two layers.

The bottom layer consists of all reagents except Taq, sterile distilled water, and template DNA. PCR was carried out in a programmable thermal controller (Mastercycler ep's Eppendorf) for 40 cycles (refer Tables 6.1 and 6.2). Amplification products were separated on a 1.4 % agarose gel, and the banding profiles were visualized in a UV transilluminator (Genei™ Imaging System, Bangalore) (Botto-Mahan et al. 2005).

Table 6.1 Reagents for RAPD analysis of reduviid predator

Constituents	Quantity	Final concentration
Sterile double-distilled water	18.5 μl	As required to make up to 25 μl
dNTPs	1 μl	200 μM of each dNTP
$MgCl_2$	1 μl	1.2 μM
10× PCR buffer	2.5 μl	1×
Primer	1 μl	50 ng
Taq polymerase	1 μl	1.5 units
Total	25 μl	

Table 6.2 Program setup for RAPD analysis

Activity	Temperature	Time	Number of cycles
Initial denaturing	94 °C	2 min	One
Denaturing	94 °C	1 min	←
Annealing	37 °C	1 min	
Extension	72 °C	1 min	← 40
Final extension	72 °C	5 min	One
Storage	4 °C	Up to end	

6.5.2 Random Amplified Polymorphic DNA (RAPD)

The dendrogram demonstrates the ability of RAPD marker for reliable differences between the diets; since phenotypic differences are not clear, know by the diets. The accessions analyzed were clustered mainly into two groups: cluster I accession of different diets with range up to 0.80 and cluster II consisting of the remaining five to six accessions with range 0.12–0.85. Lane 1 accessions are more clear than lanes 2–5. Among the six primers such as KTG-3 and KTG-5 have potentially informative value to distinguishing five diets. The presence of few polymorphic bands with less number of primers and more monomorphic bands with most primers among the five diets was observed. The presence of more number of polymorphic bands between the diets (KTG-3 and KTG-5) has separate status in *Rhynocoris marginatus*.

The present investigation of DNA profiling in diet clearly demonstrates that it was possible to analyze the RAPD pattern for correlating their similarity between the diets and accessions. On the basis of clustering data for the RAPD profile accessions of the first and second group exhibits greater among each other, and genetic variation has been found in accessions of the same populations.

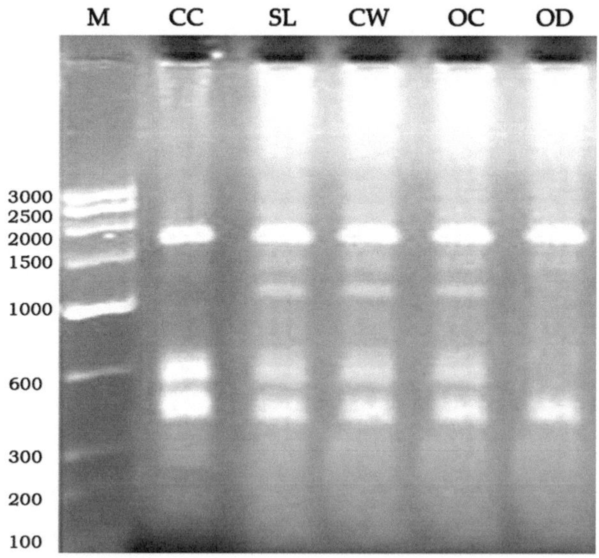

Plate 6.2 *Rhynocoris marginatus* reared with *Corcyra cephalonica* (T_1), *Corcyra cephalonica* weekly once with water (T2), oligidic diet (T3), oligidic diet and weekly once with *Corcyra cephalonica* (T_4), and *Spodoptera litura* (T_5) were subjected to RAPD analyses using KTG-3

6.5.3 Primers

Rhynocoris marginatus reared with *Corcyra cephalonica* (T_1), *Corcyra cephalonica* weekly once with water (T2), oligidic diet (T3), oligidic diet and weekly once with *Corcyra cephalonica* (T_4), and *Spodoptera litura* (T_5) were subjected to RAPD analyses using six primers. However, universal primers KTG-3 (Plate 6.2), KTG-4, KTG-5, and OPE-8 produced DNA polymorphism.

Results reveal that *Corcyra cephalonica*-reared *Rhynocoris marginatus* produced DNA polymorphism having bp ranged from 100 to 2000 with the absence of 1000, 1250, 1500, and 22,500 bp observed in other diets. *Corcyra cephalonica* weekly once with water (T_2)-, oligidic diet (T_3)-, oligidic diet and weekly once with *Corcyra cephalonica* (T_4)-, and *Spodoptera litura* (T_5)-reared *Rhynocoris marginatus* amplified the DNA of 200–2250 bp (Plate 6.3)

The DNA polymorphism with 500 bp and 700 bp (except with *Corcyra cephalonica* with KTG-4) was recorded in *Rhynocoris marginatus* reared with *Corcyra cephalonica* in KTG-5 primer. Irrespective of the prey and diets, primer DNA with bp of 200, 250, 400, 450, 600, and 2000 was common. However, none of the DNA polymorphisms with specific bp was recorded in OD. Further studies are necessary to analyze RAPD designing specific primers (Plate 6.4).

While we consider the primers, OPE-8-amplified *Rhynocoris marginatus* was fed with various diets. It showed DNA polymorphism with 450 bp was common among three diets such as *Corcyra cephalonica* weekly once with water (T_2), oligidic diet and weekly once with *Corcyra cephalonica* (T_4), and *Spodoptera litura* (T_5). Both *Corcyra cephalonica*- and OD-amplified *Rhynocoris marginatus* DNA showed specific bp such as 190 and 1500, respectively. The 190 bp DNA was not

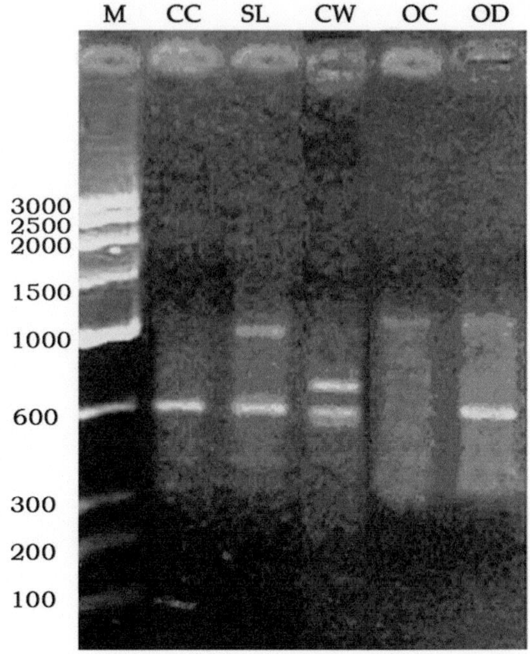

Plate 6.3 *Rhynocoris marginatus* reared with *Corcyra cephalonica* (T_1), *Corcyra cephalonica* weekly once with water (T2), oligidic diet (T3), oligidic diet and weekly once with *Corcyra cephalonica* (T_4), and *Spodoptera litura* (T_5) were subjected to RAPD analyses using KTG-4

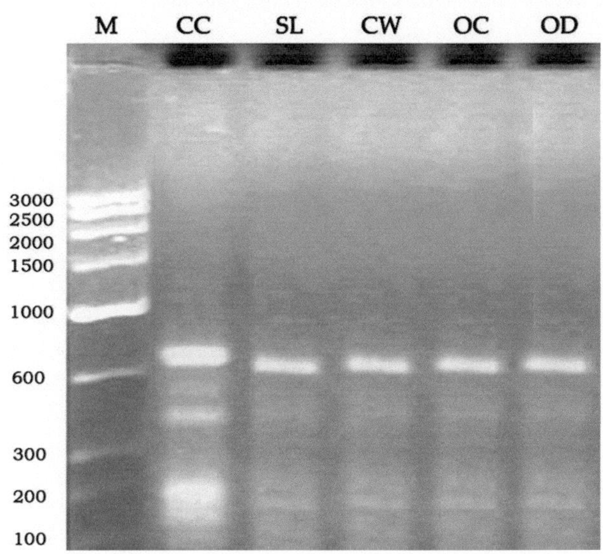

Plate 6.4 *Rhynocoris marginatus* reared with *Corcyra cephalonica* (T_1), *Corcyra cephalonica* weekly once with water (T2), oligidic diet (T3), oligidic diet and weekly once with *Corcyra cephalonica* (T_4), and *Spodoptera litura* (T_5) were subjected to RAPD analyses using KTG-5

amplified by other tested primers like KTG-3, KTG-4, and KTG-5. Similar observation was recorded for bp 1500 with other primers except KTG-3. Among the three primers, maximum and minimum DNA polymorphisms were recorded in KTG-3.

6.5.3.1 KTG-3

The resultant dendrogram drawn for *Rhynocoris marginatus* with KTG-3 primers showed all the five diets provided into two major clusters. Cluster I consisted of high similarity (0.72) obtained for oligidic diet + *Corcyra cephalonica*. Then, cluster II is divided into a, b, and c which represented for *Corcyra cephalonica* (0.68), *Spodoptera litura* (0.56), and oligidic diet. Among the five diets, *Corcyra cephalonica* harbored the lowest genetic similarity (0.25).

6.5.3.2 KTG-5

The dendrogram derived with KTG-5 revealed that genetic similarity was maximum for oligidic diet (0.82) on cluster I. For cluster II, the maximum similarity was observed at oligidic diet + weekly once with *Corcyra cephalonica*. Furthermore, cluster II is subdivided into a, b, and c with genetic similarity ranged between 0.25 and 0.78. Genetic similarity specific to KTG-5 oligidic diet shows maximum similarity, whereas cluster III gives subclusters having more or less similarity on *Corcyra cephalonica* and *Corcyra cephalonica* weekly once with water. This dendrogram shows very poor similarity value in *Spodoptera litura*.

6.5.3.3 KTG-4

The dendrogram primer KTG-4 showed that genetic similarity was maximum for oligidic diet (0.75) on cluster I. For cluster II the maximum similarity index was recorded for *Spodoptera litura* (0.72). Moreover, cluster II is subdivided into a, b, and c and their genetic similarity ranged from 0.10 to 0.63. Genetic similarity specific to KTG-4 was OD maximum similarity (0.63). As observed for KTG-3, lowest value was observed for *Corcyra cephalonica* (0.10).

6.5.3.4 OPE-8

The OPE-8, revealed that genetic similarity was high at oligidic diet in cluster I. In cluster II, high similarity was observed at oligidic diet + weekly once with *Corcyra cephalonica*. Furthermore, cluster II is subdivided into clusters a, b, and c and their genetic similarity ranged from 0.11 to 0.66. As observed for KTG-3, KTG-5, and cluster II subdivided into a, b and c, *Spodoptera litura* and *Corcyra cephalonica* + water showed minimum (0.45) and maximum (0.50) similarity index.

6.6 PCR Results

Identification of the gut contents of predatory insects can provide information on trophic relationships and the dynamics of predator–prey interactions. Several problems and constraints may be encountered in determining the diet of predatory insects in the field. Direct observations of predation in the field can be complicated by the fact that both prey and predator are often small and cryptic. Microscopic analysis of gut contents is possible for chewing predators that ingest relatively large prey fragments, but is not suitable for piercing–sucking predators.

Ingested food contains species-specific DNA sequences, thus remnant DNA provides an excellent means of detecting and identifying material from prey. Several methods, including the use of serology, are available for evaluating prey–predator relationships. Two basic methods involving precipitation and agglutination reactions have been utilized, both of which depend on the production of specific antisera with high titers. Polymerase chain reaction (PCR) detection of prey DNA has proven effective when applied to both sacrificed invertebrate predators. Considering these general opinions, we used polymerase chain reaction (PCR) for gut content analysis. It was also suggested that DNA-based methods offer the ability to identify prey where prey hard parts survive the digestion process differentially or not at all. The latter being the case for many invertebrate systems.

The present investigation on DNA profiling *Rhynocoris marginatus* clearly demonstrated that it is possible to analyze the RAPD pattern for correlating their genetic similarity and dissimilarity distance between the species within the subjected five diets regimes. Further the species specific accompany with individual diets produced the bands can be utilized define the uniqueness. Which will be a helpful in the diet to the diet change it leads to the genetic changes occurred in the predator of *Rhynocoris marginatus*.

On the basis of clustering data for the RAPD profile, accessions of the first and second group exhibit greater similarity among each other, and genetic variations have been found in accessions of the same populations because samples have been collected from the natural prey-reared predator as well as oligidic diet-maintained predators in the laboratory. Molecular profiling provides a rapid means of quantifying prey diversity within predators but when there are specific prey DNA targets, PCR with group specific primers is the principal method of choice (Symondson 2002).

Among the six primers, KTG-3 and KTG-5 have amplified better than other primers which event by amplifying; it may same sequence between the samples. The suitable primer is fine for simple laboratory studies, and when there are multiple potential target prey species (Sheppard et al. 2004) or fragments (Hoogendoorn and Heimpel 2001), the time required to assay each predator for each potential target becomes limiting. The first step in most of this research was the development of the prey-specific primer sets that were then used to amplify prey DNA from predator samples. Rapid analysis of samples is possible using this approach to study the diversity of a specific target prey range in predator gut. Generalist predators eat multiple prey species, and it is not always practical to analyze predator responses to prey diversity using the species-specific primer approach. In this study KTG-3 and KTG-5 are most preferable primers for *Rhynocoris marginatus*.

The level of genetic diversity is far greater in the oligidic diet-reared *Rhynocoris marginatus*. Although there are no available data for *Rhynocoris marginatus* or any other predatory reduviids, the logical inference is that the diversity existed in *Rhynocoris marginatus* populations. Rapid PCR-based screening systems for the study of the prey diversity of generalist predators have been developed to expand the potential of molecular detection into various areas of food web research. The techniques described by Harper et al. (2005) use a single multiplex PCR to simultaneously amplify DNA from a range of prey species.

It is extremely difficult to evaluate the biocontrol possibilities and their outcomes, when important variation in the biological control agent cannot be identified. Although the different diets (T_1, T_2, T_3, T_4, and T_5) are tested for their impact on *Rhynocoris marginatus*, critical genetic studies between populations of potential biocontrol agent. The retention time for DNA within the gut of a predator during digestion is influenced by factors including the size of the target DNA molecule. Predation by the coccinellid beetle *Coleomegilla maculata* De Geer (Coleoptera: Coccinellidae) upon the eggs of the European corn borer *Ostrinia nubilalis* (Hubner) (Lepidoptera: Crambidae) has been characterized by single-plex PCR amplification of four fragments of prey genomic DNA of different sizes from predator guts (Hoogendoorn and Heimpel 2001, 2002). Predator weight, size, developmental stage, and meal size had no effect on detection period, but in all cases the shortest fragment (150 bp) was detected for the longest time after feeding. A similar study of a coccinellid–Lepidoptera and predator–prey system in Hawaii produced analogous results with short (151 and 140 bp) prey DNA fragments being detectable in a greater proportion of beetle guts than larger (170 bp) ones (Sheppard et al. 2004).

6.7 Immunology

It is a general opinion that the development of prey-specific immunological and DNA-based techniques to detect prey in the gut contents of individual predators has been a major advance to understanding predator–prey interactions. One of these techniques showing particular promises is the enzyme-linked immunosorbent assay (ELISA) used to detect an exotic protein marker that is applied to the target prey. Mansfield et al. (2008) further developed this technique for predator–prey interactions in Australian cotton by marking *Helicoverpa armigera* eggs with rabbit IgG protein and then using the sandwich ELISA to detect its presence in two key predatory beetles: *Dicranolaius bellulus* (Guerin-Meneville) (Coleoptera: Melyridae) and *Hippodamia variegata* (Goeze) (Coleoptera: Coccinellidae). They showed that the technique was more specific and sensitive than an anti-*Helicoverpa armigera* indirect ELISA.

The works were carried out on *Lygus hesperus* (Hemiptera: Miridae) predators (Hagler et al. 1992) *Bemisia tabaci* (Homoptera: Aleyrodidae) and *P. sagitta* (Hagler and Naranjo, 1994a); *Pectinophora gossypiella* (Lepidoptera: Gelechiidae) and *P. maculiventris* (Hagler and Naranjo, 1994a, 2005); *Collops vittatus* (Coleoptera: Melyridae) (Hagler and Naranjo, 1994b), *Hippodamia convergens* (Coleoptera: Coccinellidae) (Hagler and Naranjo, 2004), *Otiorhynchus sulcatus* (Coleoptera: Curculionidae) (Crook and Solomon, 1997), *Heliothis virescens* and *Helicoverpa zea* (Lepidoptera: Noctuidae), *Orius* sp. (Heteroptera: Anthocoridae) (Sansone and Smith, 2001; Ruberson and Greenstone, 1998), and *Pterostichus melanarius* (Coleoptera: Carabidae) (Bohan et al., 2000); Araneae: Linyphiidae (Harwood et al. 2005); *Helicoverpa armigera* eggs and larvae (Lepidoptera: Noctuidae)

(Sigsgaard et al. 2002); and *Nilaparvata lugens* (Homoptera: Delphacidae) (Zhao et al., 2004; Lim and Lee 1999), Homoptera: Aphididae, *Pachygnatha degeeri* (Araneae: Tetragnathidae) (Harwood et al. 2004, 2005), Homoptera: Aphididae, and Coleoptera: Carabidae (Winder et al. 2005).

Progress in quantifying predation in agricultural systems has been hampered by the difficulty of studying predation in the field. Unlike parasitism, evidence of predation is seldom preserved in the field, and researchers must generally rely on indirect and often less precise measures of activity (Luck et al. 1988; Sunderland 1988; Naranjo and Hagler 1998). Many other factors, such as small size, nocturnal activity, cryptic behavior, and preoral digestion, contribute to the difficulty of observing and measuring predation under natural conditions (Hagler and Naranjo 1994a, b, 2005).

Techniques to study the interactions between predator and prey communities have become increasingly complex as they attempt to address the imbalance created by visual identification. These include radioisotope labeling, the application of stable isotopes, electrophoretic detection of prey isozymes, the detection of prey pigments by chromatographic analysis, and the detection of prey proteins using polyclonal antibodies (Sunderland 1988, 1996; Suntherland 2000; Pierce and Boyle 1991; Greenstone 1996; Greenstone et al. 2005).

At least a dozen different immunoassay formats have been used over the past half century to analyze arthropod guts for prey remains (Greenstone 1996); no studies were available for reduviid predators except the studies of Cohen (1993) in *Zelus renardii*. In this study, we followed indirect ELISA format for detecting factitious, natural hosts and OD remaining in adults of *Rhynocoris marginatus*.

6.7.1 Effect of Meal on Antigen and Antibody Interaction

We conducted a laboratory feeding trial to determine if *Corcyra cephalonica* (T_1), *Corcyra cephalonica* weekly once with water (T2), oligidic diet (T3), oligidic diet and weekly once with *Corcyra cephalonica* (T_4), and *Spodoptera litura* (T_5) antigen remains detectable in a predator's gut after consumption. In all instances, precautions were taken to guarantee that none of the predators collected was exposed to intermix of preys before the laboratory feeding trial. Predators were deprived of prey for 48 h prior to testing. After starving, 25 individual predators were placed into a small box (500 ml capacity) and allowed to feed one or two *Spodoptera litura*, three to four *Dysdercus cingulatus*, and three to four *Corcyra cephalonica* and 1 ml of oligidic diet separately for a minimum of hour. This is an adequate amount of time for a predator to digest food (Hagler and Cohen 1990). After each individual was observed feeding, the predator was isolated from respective food source and held for 24 h at 27 or 28 °C. Predators were removed from the respective diet and immediately for the presence of *Spodoptera litura*, *Dysdercus cingulatus*, and *Corcyra cephalonica*, and artificial diet remains in the gut of the predator were analyzed with ELISA methods. Starved predators that were not fed with respective prey and diet were also assayed (negative control). Individual predators were scored positive for the presence of prey antigen if the absorbance value was three standard deviations above that of the respective negative control mean (Hagler et al. 1993).

6.7.2 Indirect Enzyme-Linked Immunosorbent Assay (ELISA)

An indirect ELISA was performed on the predators from all three feeding trials. A 100 μl aliquot of each macerated predator was placed in an individual well of a 96-well assay plate (round-bottomed wells of polystyrene microtiter plates, Nunc MaxiSorp, UK). Each plate was incubated at 4 °C overnight. Following incubation, the insect macerates were discarded from each plate, and free reactive sites were blocked with 1 % bovine serum albumin (BSA) in distilled water (blocking solution) which was added to each well for 30 min to block any unoccupied antigenic sites in the wells. The blocking solution was emptied from each plate, and a 50 μl aliquot of *Spodoptera litura*, *Dysdercus cingulatus*, *Corcyra cephalonica*, and pig liver was added with rabbit monoclonal antibody (MAb) using acetic fluid. It was mixed with diluted 1:10,000 BSA and was added to each well of the ELISA plate (Hagler et al. 1994).

The ELISA plates were then incubated for 1 h at room temperature. The contents from each plate were discarded, and the plates were briefly rinsed three times with PBS-Tween 20 (0.05 %) and twice with phosphate-buffered saline (PBS). Goat anti-mouse IgG conjugated to alkaline phosphatase diluted (1:500) in blocking solution (BSA) was added to each well (50 μl) of the plates for 1 h. Plate contents were discarded, and again plates were rinsed three times as described above. A 50 μl aliquot of substrate solution was added to each well using the reagents supplied in an alkaline phosphatase substrate kit (Nunc MaxiSorp, UK) following the addition of 50 μl of 2 N H_2SO_4. After 4 h, the absorbance of each well was measured with a SLR 36 ELISA microplate reader at 450 nm.

6.7.3 Effect of Predator Protein Content on ELISA Sensitivity

Protein content of *Rhynocoris marginatus* in relation to three pests *Spodoptera litura*, *Dysdercus cingulatus*, and *Corcyra cephalonica* and pig liver-based oligidic diet is significantly different. The overall mean protein concentration was 400, 330, 400, 475, and 425 μg/insect, respectively. After the determination of the total protein content, each predator was assayed by a standardized indirect ELISA in which each predator was homogenized, and its respective prey antigen was detected in each individual predator sample that was spiked with a single pest, yielding a mean ELISA absorbance value. Indirect ELISA was examined to detect the influence of *Spodoptera litura*, *Corcyra cephalonica*, and oligidic diet numerous in adult *Rhynocoris marginatus*. *Rhynocoris marginatus* samples were homogenized in 1500 μl PBS, and then each sample was analyzed by an indirect ELISA.

Of the predator species fed, *Spodoptera litura*- and OD-reared predators yielded consistently high (positive) absorbance values (Fig. 6.2) followed by *Corcyra cephalonica*-fed predator. There is no significant difference observed in *Corcyra cephalonica*-fed predator. The results from this study showed that the sensitivity of the indirect ELISA depended on the prey species and quantity of prey consumed by predator. The more preys eaten by a given predator species, the greater the probability of obtaining a positive ELISA reaction. The result of this study revealed that the sensitivity of the indirect ELISA depended on the prey species and quantity of the prey consumed by the predator. Predator digestion rate, prey size, predator size, diet, and the physiological state of the prey can affect the outcome of a gut content immunoassay (Sunderland 1996). Before a precise estimate of predation can be made, these factors must be considered. Despite these characteristics, immunoassays remain a qualitative method that provides direct evidence of predation by specific species but rarely provides quantitative estimates of the number of prey killed. However, previous studies (Sopp and Sunderland 1989; Symondson and Liddell 1993; Shapiro and Legaspi 2006), suggest that gut content ELISAs vary in efficacy among pest species. One of the fundamental parameters for qualitatively or quantitatively estimating predation using immunoassays is the period of time the prey antigens remain detectable in a predator's gut. The detection interval is a key parameter in most indices that have been developed to assess predation using immunoassays (Sopp et al. 1992) and is very important in comparative evaluations of different predator species feeding on the same prey (Cohen 1998b). The more quantity of prey eaten by a given predator species, the greater the probability of obtaining a positive ELISA reaction. The indirect ELISA is suitable for detecting prey remains in the gut of *Rhynocoris marginatus*. Only those individuals consumed by *S. litura* or all lepidopteron pests. *Corcyra cephalonica*-reared predators are also showing positive but not like *S. litura*.

The overall mean protein concentration (± SE) was high in *Spodoptera litura* followed by *Rhynocoris marginatus*, *Dysdercus cingulatus* and pig liver, and *Corcyra cephalonica*. This shows a strong linear relationship between insect weight and total protein. This suggests that weight alone can be used to estimate the total protein content of a predator. Therefore, the optimal quantity of PBS that a given predator species showed to be homogenized can be determined simply by weighing the predators (says Hagler et al. 1997). They further reported that using the weight of an animal to predict its protein content eliminates labor-intensive protein analyses and chemical waste. *Spodoptera litura* have the highest proportion followed by artificial diet reared predator has highest proportion of yield of immunoreactions, suggesting that they are effective at detecting the presence of a small quantity of prey in whole body homogenized predators in examined in the feeding trial were very effective at detecting and *Dysdercus cingulatus*, *Spodoptera litura*, *Corcyra cephalonica* and artificial diet reared predators. *Spodoptera litura*, *Corcyra cephalonica* are under order Lepidoptera if feed up on leaves chewing type, while *Dysdercus cingulatus* are feed plant sap by sucking so body physiological conditions also may entirely different. So this may the reason be for the variation among the diets too.

Current predator–prey studies, however, tend to rely on monoclonal antibody and/or DNA-based technology, which allow accurate and rapid detection of prey remains within predator guts or fecal samples. Immunological assays using prey-specific protein antibodies (Greenstone and Hunt 1993; Powell et al. 1996; Hagler and Naranjo 1997; Hagler 1998; Symondson et al. 1999a, b; Shapiro and Legaspi 2006) was widely used to identify predator gut contents. These assays are used to determine the absence or presence of prey in the gut. The accuracy of the assay depends on several factors, including temperature, meal size, time since feeding, resistance of the target protein to digestion, and predator species (Hagler and Naranjo 1997; Symondson 2002; Symondson et al. 1997, 1999b; Hagler 1998).

In the context of biological control, increased prey biodiversity would be predicted to enhance the ability of populations of generalist predators to achieve sustainable levels of pest control. However, pest species are often poor on terms of nutritional quality (Toft 1999, 2005) and are avoided by some predators (Toft 1997). Hence, increased dietary diversity has the potential of the predators from feeding on pests and ultimately reduces levels of biological control (Halaj and Wise 2002; Madsen et al. 2004; Koss and Snyder 2005; Wise et al. 2006). Elucidating possible shifting preference of oligidic diet (OD) for alternative non-pest prey in complex food webs and uncovering the strength of those tropic linkages are therefore essential for incorporating biological control by predators reared with oligidic diet in integrated pest management (IPM).

Of the many methods used for studying predation, postmortem approaches are among the most direct and least likely to introduce bias through unintentional experimental disruption (Luck et al. 1988; Sunderland 1988). Postmortem methods include gut dissection and chromatographic, electrophoretic, PCR, and immunological analysis of predator gut contents. Depending on the type of antibody and assay system used, immunological methods can be species or stage specific, highly sensitive, and rapid enough to facilitate screening of thousands of predators in a shortest period (Sunderland 1988; Greenstone 1996) (Fig. 6.2).

Variable predator digestive rates (Symondson and Liddell 1993), predator–prey size (Sopp and Sundarland 1989), temperature (Mclver 1981), predator metabolic status (Lovei et al. 1990), and the developmental stage of the prey (Hagler et al. 1992) can affect the quantitative outcome of gut content immunoassays (Sundarland 1996). The data presented have suggested that there is a huge discrepancy in the sensitivity of a gut content immunoassay developed to detect *Spodoptera litura* and *Dysdercus cingulatus*; *Corcyra cephalonica* remains in the whole homogenized predators. Despite these complications gut content immunoassays after unique opportunities to examine the natural feeding behavior of predators from unmanipulated systems, and after advantage of permitting thousands of predators to be assayed rapidly and inexpensively (Hagler and Naranjo 1994a, b, 2005; Sheppard and Harwood 2005; Winder et al. 2005).

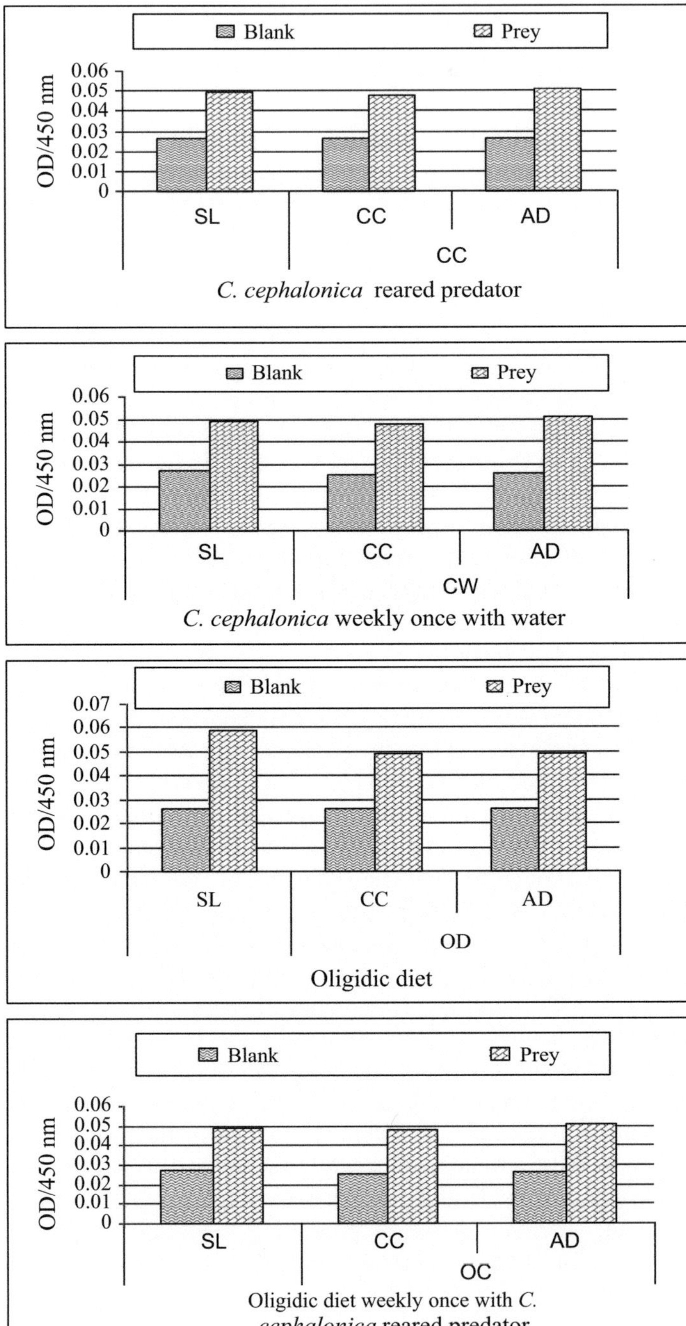

Fig. 6.2 Titers of antibody in indirect ELISA with pest antigen from *Rhynocoris marginatus* gut at different diets

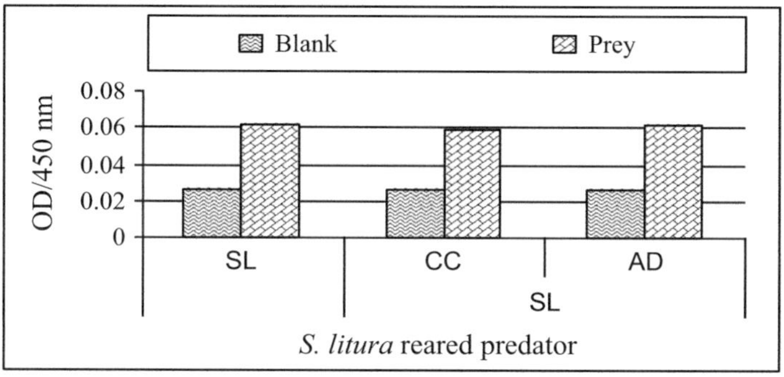

Fig. 6.2 (continued)

6.8 Conclusions

Carbohydrate, protein, and lipid content in relation to various foods was provided to the reduviid. Indirect ELISA was used to detect the influence of *Spodoptera litura*, *Corcyra cephalonica*, and OD in adult *Rhynocoris marginatus*. The *Spodoptera litura* (0.062)- and OD-reared predators yield the highest frequency of positive immunoassay followed by *Corcyra cephalonica*-reared predator. It is not always possible to track trophic interactions between predators and prey by direct observation. Hence an advanced technology with PCR and ELISA were used in this study.

Six primers KTG-3, KTG-5, KTG-4, OPE-8, OPE-13, and OPE-16 were tested to amplify the DNA of the predators' population. The size of the amplified DNA ranged from 200 to 2000 bp. A total of 52 amplicons resulted from four primers in five populations of *Rhynocoris marginatus* analyzed. The number of amplifications was 24, 16, and 7 for KTG-3, KTG-5, and KTG-4, respectively. The lowest number of amplicon polymorphisms was produced with the primer OPE-8. The electrophoretic pattern reveals polymorphism was present among *Rhynocoris marginatus* populations.

6.9 Future Recommendations

- Protein profiling will be carried out in fat bodies after being fed with various diets and also field-collected reduviid predators.
- Sandwich ELISAs will be utilized to detect the presence of exotic protein markers which are very sensitive and are an effective way of identifying potential predators of marked prey.

References

Agusti N, Cohen AC (2000) *Lygus hesperus* and *L.lineolaris* (Hemiptera: Miridae), phytophages, zoophages, or omnivores: evidence of feeding adaptations suggested by the salivary and midgut digestive enzymes. J Entomol Sci 35:176–186

Agusti N, de Vicente MC, Gabarra R (1999) Development of sequence amplified characterized region (SCAR) markers of *Helicoverpa armigera*: a new polymerase chain reaction-based technique for predator gut analysis. Mol Ecol 8:1467–1474

Agusti N, de Vicente MC, Gabarra R (2000) Developing SCAR markers to study predation on *Trialeurodes vaporariorum*. Insect Mol Biol 9:263–268

Aussel J-P, Linley JR (1994) Natural food and feeding behavior of *Culicoides furens* larvae (Diptera: Ceratopogonidae). J Med Entomol 31:99–104

Bohan DA, Bohan AC, Glen DM, Symondson WOC, Wiltshire CW, Hughes L (2000) Spatial dynamics of predation by carabid beetles on slugs. J Anim Ecol 69:367–379

Bradfield JY, Berlin RL, Keeley LL (1990) Contrasting modulation of gene expression by a juvenile hormone analog. Insect Biochem 20:105–111

Botto-Mahan C, Ortiz S, Rozas M, Cattan PE, Solri A (2005) DNA evidence of *Trypanosoma cruxi* in the Chilean vector *Mepraria spinolai* (Hemiptera: Reduviiae). Mem Insect Oswalde Crux Rio de Ianeire 100(3):237–239

Chen Y, Giles KL, Payton ME, Greenstone MH (2000) Identifying key cereal aphid predators by moleculargut analysis. Mol Ecol 9:1887–1898

Cohen AC (1993) Organization of digestion and preliminary characterization of salivary trypsin-like enzymes in a predaceousheteropteran, *Zelus renardii*. J Insect Physiol 39:823–829

Cohen AC (1998a) Solid-to-liquid feeding: the inside(s) story of extra-oral digestion in predaceous Arthropoda. Am Entomol 44:103–117

Cohen AC (1998b) Biochemical and morphological dynamicsand predatory feeding habitsin terrestrial heteroptera. In: Ruberson JR, Coll M (eds) Predaceous heteroptera: implications for biological control. Thomas Say Publications in Entomology. Entomological Society of America, Lanham, pp 21–32

Crook A, Solomon M (1997) Predators of vine weevil in soft fruit plantations. New developments in the soft fruit industry. Proceedings of a conference organized by ADAS, HRI and EMRA in Association with CCFRA, HDC and the Grower Magazine, pp 83–87. Horticulture Research International, Ashford, UK

Dodd C, Bruford M, Symondson W (2003) Detection of slug DNA within carabid (Coleoptera: Carabidae). Annu Rev Entomol 41:231–256

Gordon ER, Weirauch C (2015) Efficient capture of natural history data reveals prey conservatism of cryptic termite predators. Mol Phylogenet Evol 94:65–73

Greenstone MH (1996) Serological analysis of arthropod predation: past, present and future. In: Symondson WOC, Liddell JE (eds) The ecology of agricultural pests: biochemical approaches. Chapman and Hall, London, pp 265–300

Greenstone MH, Hunt JH (1993) Determination of prey antigen half-life in *Polistes metricus* using a monoclonal antibody-based immunodot assay. Entomol Exp Appl 68:1–7

Greenstone MH, Rowley DL, Heimbach U, Lundgren JG, Pfannenstiel RS, Rehner SA (2005) Barcoding generalist predators by polymerase chain reaction: carabids and spiders. Mol Ecol 14:3247–3266

Greenstone MH, Tillman PG, Hu JS (2014) Predation of the newly invasive pest *Megacopta cribraria* (Hemiptera: Plataspidae) in soybean habitats adjacent to cotton by a complex of predators. J Econ Entomol 107(3):947–954

Hagler JR (1998) Variation in the efficacy of several predator gut content immunoassays. J Biol Control 12:25–32

Hagler JR, Cohen AC (1990) Effects of time and temperature on digestion of purified antigen by *Geocoris punctipes* (Hemiptera: Lygaeidae) reared on artificial diet. Ann Entomol Soc Am 83:1177–1180

Hagler JR, Naranjo SE (1994a) Determining the frequency of heteropteran predation on sweetpotato whitefly and pinkbollworm using multiple ELISAs. Entomol Exp Appl 72:59–66

Hagler JR, Naranjo SE (1994b) Qualitative survey of two coleopteran predators of *Bemisia tabaci* (Homoptera: Aleyrodidae) and *Pectinophora gossypiella* (Lepidoptera: Gelechiidae) using multiple prey gut content ELISA. Environ Entomol 23:193–197

Hagler JR, Naranjo SE (1997) Measuring the sensitivity of an indirect predator gut content ELISA: detectability of prey remains in relation to predator species, temperature, time, and meal size. J Biol Control 9:112–119

Hagler JR, Naranjo SE (2004) A multiple ELISA system for simultaneously monitoring intercrop movement and feeding activity of massreleased insect predators. Int J Pest Manag 50:199–207

Hagler JR, Naranjo SE (2005) Use of a gut content ELISA to detect whitefly predator feeding activity after field exposure to different insecticide treatments. Biocontrol Sci Tech 15:321–339

Hagler JR, Cohen AC, Bradley-Dunlop D, Enriquez FJ (1992) Field evaluation of predation using a species-andstage-specific monoclonal antibody. Environ Entomol 21:896–900

Hagler JR, Brower AG, Tu Z, Byrne DN, Bradley-Dunlop D, Enriquez FJ (1993) Development of a monoclonal antibody to detect predation of the sweetpotato whitefly, *Bemisia tabaci*. Entomol Exp Appl 68(3):231–236

Hagler JR, Naranjo SE, Bradley-Dunlop D, Enriquez FJ, Henneberry TJ (1994) A monoclonal antibody to pink bollworm (Lepidoptera: Gelechiidae) egg antigen: a tool for predator gut analysis. Ann Entomol Soc Am 87(1):85–90

Hagler JR, Naranjo SE, Erickson ML, Machtley SA, Wright SF (1997) Immunological examinations of species variability in predator gut content assays: effects of predator: prey protein ratio on immunoassay sensitivity. J Biol Control 9:120–128

Halaj J, Wise DH (2002) Impact of a detrital subsidy on trophic cascades in a terrestrial grazing food web. Ecology 83:3141–3151

Harper GL, King RA, Dodd CS, Harwood JD, Glen DM, Bruford MW, Symondson WOC (2005) Rapid screening of invertebrate predators for multiple prey DNA targets. Mol Ecol 14:819–828

Harwood JD, Sunderland KD, Symondson WOC (2004) Prey selection by linyphiid spiders: molecular tracking of the effects of alternative prey on rates of aphid con-sumption in the field. Mol Ecol 13:3549–3560

Harwood JD, Sunderland KD, Symondson WOC (2005) Monoclonal antibodies reveal the potential of the tetragnathid spider *Pachygnatha degeeri* (Araneae: Tetragnathidae) as an aphid predator. Bull Entomol Res 95:161–167

Hoogendoorn M, Heimpel GE (2001) PCR-based gut content analysis of insect predators: using ribosomal ITS-1 fragments from prey to estimate predation frequency. Mol Ecol 10:2059–2067

Hoogendoorn M, Heimpel GE (2002) PCR-based gut content analysis of insect predators: using ITS-1 fragments from prey to estimate predation frequency. In: Van Dreische RG (ed) Proceedings of the 1st international symposium on biological control of arthropods, Honolulu, Hawaii, 2002. USDA, Morgantown, pp 91–97

Jeyaraman J (1981) Laboratory manual in biochemistry. Willey Limited, New Delhi, p 85

Kajiura Z, Yamashita O (1989) Stimulated synthesis of the female specific staorage protein in male larvae of the silkworm, *Bombyx mori*, treated with junvenile hormone analog. Arch Insect Biochem Physiol 12:99–109

Koss AM, Snyder WE (2005) Alternative prey disrupt biocontrol by a guild of generalist predators. Biol Control 32:243–251

Laemmli UK (1970) Cleavage of structural proteins during the assembly of the head of bacteriophage T_4. Nature 227:680–685

Legaspi JC, Legaspi BC, Meaghher RL, Ciomperlik MA (1996) Evaluation of *Serangium parcesetosum* (Coleoptera: Coccinellidae) as a biological control agent of the silverleaf whitefly (Homoptera: Aleyrodidae). Environ Entomol 25:1421–1427

Lim UT, Lee JH (1999) Enzyme-linked immunosorbent assay used to analyze predation of *Nilaparvata lugens* (Homoptera: Delphacidae) by *Pirata subpiraticus* (Araneae: Lycosidae). Environ Entomol 28:1177–1182

Lovei GL, Sopp PI, Sunderland KD (1990) Digestion rate in relation to alternative feeding in three species of polyphagouus predators. Ecol Entomol 15:293–300

Luck RF, Shepard BM, Kenmore PE (1988) Experimental methods for evaluating arthropod natural enemies. Annu Rev Entomol 33:367–391

Madsen M, Terkildsen S, Toft S (2004) Microcosm studies on control of aphids by generalist arthropod predators: effects of alternative prey. BioControl 49:489–504

Mansfield S, Hagler JR, Whitehouse MEA (2008) A comparative study of the efficiency of a pest-specific and prey-marking ELISA for detection of predation. Entomol Exp Appl 127:199–206

McIver J (1981) An examination of the utility of the precipitin test for evaluation of arthropod predator-prey relationships. Can Entomol 113:213–222

Mjeni AM, Morrison PE (1976) Junvenile horomoe analogue and egge development in the blowfly, *Phormia regina* (Meig.). Gen Comp Endocrinol 28:321–349

Morales ME, Wesson DM, Sutherland IW, Impoinvil DE, Mbogo CM, Githure JI, Beier JC (2003) Determination of *Anopholes gamibiae* larval DNA in the gut of insectivorous dragonfly (Libellulidae) nymphs by polymerase chain reaction. J Am Mosq Control Assoc 19:163–165

Munyaneza J, Obrycki JJ (1998) Searching behavior of *Coleomegilla maculata* larvae feeding on Colorado potato beetle eggs. Biol Control 13:85–90

Naranjo SE, Hagler JR (1998) Characterizing and estimating the impact of heteropteran predation. In: Coll M, Ruberson J (eds) Predatory Heteroptera: their ecology and use in biological control. Entomological Society of America, Lanham, pp 170–197

Nigam CS, Omkar (2003) Experimental animal physiology and biochemistry. New Age International (p) Limited, Publishers, New Delhi

Pierce GJ, Boyle PR (1991) A review of methods for dietanalysis in piscivorous marine mammals. Oceanogr Mar Biol 29:409–486

Powell W, Walton MP, Jervis MA (1996) Populations and communities. In: Jervis M, Kidd NAC (eds) Insect natural enemies. Chapman and Hall, London, pp 223–293

Ruberson JR, Greenstone MH (1998) Predators of budworm/bollworm eggs in cotton: an immunological study. Proceedings of the 1998 Beltwide Cotton Conferences, 2: 1095–1098

Sahayaraj K (2004) Reduviids in biological control. In: Sahayaraj K (ed) Indian insect predators in biological control. Dayas Publication, India, pp 134–166. ISBN 8170353408

Sambrook J, Fritsch EF, Manistis T (1989) In: Nolan C (ed) Molecular cloning: a laboratory manual volumes I – III. Cold Spring Harbor Laboratory Press, Cold Spring Harbor

Sansone CG, Smith JW Jr (2001) Identifying predation of *Helicoverpa zea* (Lepidoptera: Noctuidae) eggs by *Orius* spp. (Hemiptera: Anthocoridae) in cotton by using ELISA. Environ Entomol 30:431–438

Shapiro JP, Legaspi JC (2006) Assessing biochemical fitness of predator *Podisus maculiventris* (Heteroptera: Pentatomidae) in relation to food quality: effects of Wve species of prey. Ann Entomol Soc Am 99:321–326

Sheppard SK, Harwood JD (2005) Advances in molecular ecology: tracking trophic links through predator–prey food-webs. Funct Ecol 19(5):751–762

Sheppard SK, Henneman ML, Memmott J, Symondson WOC (2004) Infiltration by alien predators into invertebrate food webs in Hawaii: a molecular approach. Mol Ecol 13:2077–2088

Sigsgaard L, Greenstone MH, Duffield SJ (2002) Egg cannibalism in *Helicoverpa zea* on sorghum and pigeonpea. Biol Control 47:151–165

Sleaford F, Bignell DE, Eggleton P (1996) A pilot analysis of gut contents in termites from the Mbalmayo Forest Reserve, Cameroon. Ecol Entomol 21:279–288

Sopp PI, Sunderland KD (1989) Some factors affecting the detection period of aphid remains in predators using ELISA. Entomol Exp Appl 51:11–20

Sopp PI, Sunderland KD, Fenlon JS, Wratten SD (1992) An improved quantitative method for estimating invertebrate predation in the field using an enzyme-linked immunosorbent assay (ELISA). J Appl Ecol 29:295–302

Sunderland KD (1988) Quantitative methods for detecting invertebrate predation occurring in the field. Annu Appl Biol 112:201–224

Sunderland KD (1996) Progress in quantifying predationusing antibody techniques. In: Symondson WOC, Liddell JE (eds) The ecology of agricultural pests: biochemical approaches. Chapman and Hall, London, pp 419–455

Suntherland RM (2000) Molecular analysis of avian diets. PhD Thesis, University of Oxford, Oxford

Symondson WOC (2002) Molecular identification of preying predator diets. Mol Ecol 11:627–641

Symondson WOC, Liddell JE (1993) A monoclonal antibody for the detection of arionid slug remains in carabid predators. Biol Control 3:207–214

Symondson WOC, Erickson ML, Liddell JE (1997) Species-specific detection of predation by coleoptera on the milacid slug *Tandonia budapestensis* (Mollusca: Pulmonata). Biocontrol Sci Tech 7:457–465

Symondson WOC, Erickson ML, Liddell JE (1999a) Development of a monoclonal antibody for the detection and quantification of predation on slugs within the *Arion hortensis* Agg. (Mollusca: Pulmonata). Biol Control 16:274–282

Symondson WOC, Erickson ML, Liddell JE, Jayawardena KGI (1999b) Amplified detection, using a monoclonal antibody, of an aphid-specific epitope exposed during digestion in the gut of a predator. Insect Biochem Mol Biol 29:873–882

Toft S (1997) Acquired food aversion of a wolf spider to three cereal aphids: intra- and interspecific effects. Entomophaga 42:63–69

Toft S (1999) Prey choice and spider fitness. J Arachnol 27:301–307

Toft S (2005) The quality of aphids as food for generalist predators: implications for natural control of aphids. Eur J Entomol 102:371–383

Triltsch H (1997) Gut contents in field sampled adults of *Coccinella septempunctata* (Coleoptera: Coccinellidae). Entomophaga 42:125–131

Winder L, Alexander CL, Holland JM, Symondson WOC, Perry J, Woolley C (2005) Predatory activity and spatial pattern: the response of generalist carabids to their aphid prey. J Anim Ecol 74:443–454

Wise DH, Moldenhauer DM, Halaj J (2006) Using stable isotopes to reveal shifts in prey consumption by generalist predators. Ecol Appl 16:865–876

Zaidi RH, Jaal Z, Hawkes NJ, Hemingway J, Symondson WOC (1999) Can multiplecopy sequences of prey DNA be detected amongst the gut contents of invertebrate predators? Mol Ecol 8:2081–2087

Zhao WC, Cheng JA, Chen ZX (2004) Development of a monoclonal antibody to detect predation of the brown planthopper *Nilaparvata lugens* (Stal). Int J Pest Manag 50:317–321

Field Evaluation

7

Abstract

A total of five pests such as *Aphis craccivora*, *Helicoverpa armigera*, *Spodoptera litura*, *Mylabris indica*, and grasshoppers were recorded both in summer and kharif. However, these pest populations were drastically decreased after introduction of oligidic diet-reared predator. Among all predators recorded, *Rhynocoris marginatus* was the most abundant in the groundnut field. From this result, it was very clear that *Rhynocoris marginatus* greatly suppressed the population of *Spodoptera litura*, *Aphis craccivora*, *Helicoverpa armigera*, and *Pericallia ricini*. In summer *Rhynocoris fuscipes*, spiders, dragonfly, *Mantis religiosa*, and *Coccinella septempunctata* were established. Among the two-diet regime, OD-reared *Rhynocoris marginatus* slightly and highly reduce the pest populations.

Production of the groundnut was high in the artificial diet-reared predator. Similarly the cost-benefit ratio was maximum in the field-released OD predator (1:2) followed by T_1 (CC)-reared predator (1:1.8). Among the treatments, number of two pods was more than three pods in a groundnut plant. During summer, groundnut production was maximum in the oligidic diet-reared predator plot (1224 kg/h) followed by T_1 (936.00 kg/h) and minimum in the control plot (728.00 kg/h). Same trend was also observed in kharif. During kharif, the percent avoidable loss (PAL) was maximum (23.33 %) in field-released OD-reared predator, followed by T_1-reared predator released (14.81 %).

Keywords

Integrated pest management • *Zelus* • *Coranus* • *Rhynocoris* • *Pristhesancus* • Economically imporant crop • Cost-benefit ratio

K. Sahayaraj, R. Balasubramanian, *Artificial Rearing of Reduviid Predators for Pest Management*, DOI 10.1007/978-981-10-2522-8_7

7.1 Biological Control: An Overview

Biological control is often viewed as a promising alternative or complement to pesticides in integrated pest management (IPM) programs (Bengtsson et al. 2005; Hole et al. 2005). The successes and failures of biological control have been extensively reviewed (van Driesche and Bellows 1996; Cardinale et al. 2003; Aquilino et al. 2005; Byrnes et al. 2005; Wilby et al. 2005). Factors that can influence the effectiveness of biological control agents include agent specificity (generalist or specialist), the type of agent (predator, parasitoid, or pathogen), the timing and number of releases, the method of release, synchrony of the natural enemy with the host, field conditions, and release rate. Augmentative, or inundative, biological control is the release of large numbers of natural enemies to augment natural enemy populations or inundate pest populations with natural enemies (Snyder et al. 2008).

7.2 Commercial Available Entomophagous Insects

More than 125 species of entomophagous arthropods are commercially used as augmentative biological control. Augmentative release of the predator is a main component in the IPM, and especially the reduviids play a major role in the suppression of various pests of economic importance (Schaeffer 1988; Sahayaraj 1999a, b, 2002b; Sahayaraj and Martin 2003, Sahayaraj and Ravi 2007a, b).

Before introducing any biocontrol agents, it is imperative to evaluate under field-cage or pot conditions. For instance *Arilus cristatus* (L.) is consumed by most medium-sized larvae of the soybean looper, *Pseudoplusia includens* (Lepidoptera: Noctuidae) (Richman et al. 1980). *Scipinia horrid* (Stal) is considered as a good biocontrol agent of *Heteropsylla cubana* Crawford in Thailand and also in Mexico (Napompeth et al. 1989). Recently *Zelus obscuridorsis* (Hemiptera: Reduviidae) was recorded as a predator of the South American tomato leaf miner, *Tuta absoluta* (Lepidoptera: Gelechiidae) (Speranza et al. 2014).

More than 120 pests attack groundnut both at crop stage and storage (Ramaraju et al. 1998). The pests are classified as defoliators, borers, and sucking pests. The important defoliators are the larval forms of *Amsacta albistriga* Walker, *Aproaerema modicella* Dev., *Spodoptera litura* Fab., *Helicoverpa armigera* Hübner, etc. The cause of severe damage to groundnut (Amin 1983; Panchabhavi and Nethradhaniraj 1987; Wightman and Rao 1993). The irrigated groundnut in Tamil Nadu is severely attacked by *Spodoptera litura* and *Helicoverpa armigera* (Peter and David, 1998). The other major pests of groundnut are jassids (Biswas et al. 2000), thrips, white flies, bugs, beetles, and grasshoppers (Jayanthi et al. 2000; Sridhar and Mahto 2000). The extent losses of groundnut are feeding, and transmitting virus disease of aphids is well reported (Usman et al. 2015). Grasshopper causes leaf damage and yield loss in groundnut. It was controlled by both biological and chemical agents in groundnut field (Peveling et al. 1999). However, no predatory bugs are available for groundnut pest management.

7.3 Field Evaluation of Reduviids: A Worldwide Scenario

7.3.1 USA

Zelus Fabricius is a widespread genus across the whole of the Americas, with about 60 described species. It preys on a wide variety of insects, including lacewings, Hymenoptera, small Coleoptera, Diptera, Hemiptera, and other small arthropods. *Zelus* has been known to reach extremely large population densities, as much as 50,000–75,000 per hectare (Ables 1978). Though nine species of *Zelus* Fabricius, 1803, are found in the United States, Canada, and northern Mexico, *Zelus tetracanthus* Stal, *Zelus luridus* Stal, *Zelus exsanguis* Stal, *Zelus nugax* Stal, *Zelus renardii* Kolenati, *Zelus cervicalis* Stal, *Zelus janus* Stal, *Zelus longipes* (L.), and *Zelus grassans* Stal, however, none is utilized in the biological control program (Hart 1986).

Similarly, *Sinea* is a genus of relatively small, brown, New World harpactocorine assassin bugs, which occurs from across Canada, southern United States, Mexico, and the Caribbean to northern South America [*Sinea diadema*, *Sinea confuse*]; none has been utilized in IPM or biocontrol program so far. Another important reduviid recorded in the United States is wheel bug, *Arilus cristatus* (Linnaeus). The wheel bug occurs throughout Florida. It has been reported from Rhode Island west to California and south to Texas and Florida. Blatchley (1926) included Mexico and Guatemala in range. Wygodzinsky (1949) recognized four species of *Arilus* in this New World genus, but only *Arilus cristatus* occurs in the United States. The wheel bug has been reported preying upon a wide variety of insects in several orders including Lepidoptera, especially Arctiidae and Noctuidae; Coleoptera, especially Chrysomelidae and Coccinellidae; Hemiptera, especially Pentatomidae; Hymenoptera, especially sawflies; and Homoptera, especially aphids upon which young nymphs feed. Unfortunately, the above groups include the beneficial lady beetles and honey bees. A distinguishing feature is a large spine dorsally on the fore-femora (Mead 1999).

Approximately 200 specimens of predators representing 16 species were evaluated in field-cage tests to determine daily consumption levels of eggs and larvae of the soybean looper, *Pseudoplusia includes* (Walker). The highest predation rates were recorded for the nabids *Reduviolus roseipennis* (Reuter), *Tropiconabis capsiformis* (Germar), and *Hoplistoscelis deceptivus* (Harris); the green lacewing, *Chrysopa rufilabris* (Burmeister); and the lygaeids *Geocoris punctipes* (Say) and *Geocoris uliginosus* (Say). Consumption of small larvae was largest for the nabids *Tropiconabis capsiformis* and *Hoplistoscelis deceptivus*, the lygaeids *Geocoris punctipes* and *Geocoris uliginosus*, and the carabid beetle *Calleida decora* (F.). The pentatomid *Stiretrus anchorago* (F.) and the reduviid *Arilus cristatus* (L.) consumed the most medium-sized larvae (Richman et al. 1980).

7.3.2 Egypt

In Egypt, *Coranus africana* El-Sebaey was considered as a biological control agent against *Spodoptera littoralis* Bosid and *Agrotis ipsilon* Rott and utilized to evaluate suppression of efficiency against white fly *Bemisia tabaci* in tomato field (El-Sebaey et al. 2004). Results showed that the reduction percentage of *Bemisia tabaci* infestation differed according to the release of predator in each treatment. For instance, during the first year in the case of release of the adult stage, the reduction in the second week of release was 93.4 % and 100 % for adult and immature stage. During the second year, the reduction was 94.8 % and 97.1 %. Previously, El-Sebaey et al. (2003), tried in tomato field to manage *Bemisia tabaci* (Genn.) (El-Sebaey et al., 2011), *Aphis gossypii* Glov., and *Spodoptera littoralis* (Bosid). Later, the same reduviid was used to control *Aphis gossypii* Glov. in cucumber and squash fields (El-Sebaey et al. 2007) and also to manage *Bemisia tabaci* with augmentative release in cucumber greenhouse (El- Sebaey et al. 2002).

Mass-reared reduviid, *Coranus africana* El-Sebaey, was released in a cucumber greenhouse (500 m^2) against *Bemisia tabaci* (Genn.). A greenhouse without the predator was considered as control during the study period. After the first release, *Coranus africana* reduced the infestation of *Bemisia tabaci* by 79.7 % and 82.6 % for adult and immature stage, respectively, on the 55th day after sowing. The reduction was increased after the second release (85th day after sowing) to reach 97.7 % and 96.4 % for adult and immature stages, respectively. The reduction was 81 % and 80.36 % on the 55th and 85th day after sowing and 98.7 % and 96.2 % for adult and immature stages, respectively, during the second year. The early and total yield of cucumber were increased in the first year from 0.77 to 1.26 and from 1.11 to 1.98 kg. of fruit/plant, respectively. The correspondence value for the second year followed the same trend from 0.84 to 1.4 and from 1.2 to 2.03 kg. of fruit/plant (http://enbii.cu.edu.eg/Pests/IPM/paper_8_27.pdf) (El- Sebaey et al. 2002).

Also the evolution of the assassin bug *Coranus africana* El-Sebaey was studied in the suppression of different populations of *Aphis gossypii* in cucumber and squash field at Faiyum Governorate during the successive years (2002–2003). *Coranus africana* was released in the ratio of one predator/plant against three levels of *Aphis gossypii* infestations in cucumber and squash field. Control experimental plants without *Coranus africana* were maintained for each level during the evaluation period. Reduction rates of *Aphis gossypii* infestation differed according to the number of predator in each treatment. In case of high infestation, the reduction was 99.9 % and 92.86 % during the first and second year in cucumber field as opposed to 88.6 % and 90.1 % in squash field. A complete eradication of the pest population was achieved in the case of low and moderate infestation levels in two crops. The basic yield parameters expressed as the weight of fruit and number of fruits during 2 years in the two crops (http://enbii.cu.edu.eg/Pests/IPM/paper_8_27.pdf).

7.3.3 India

7.3.3.1 Coconut

In India, an exotic reduviid predator *Platymeris laevicollis* was colonized and the laboratory-reared bugs were released in large numbers on the coconut palm in Kerala, Lakshadweep, and Karnataka, and they controlled the beetle *Oryctes rhinoceros* Linn (Antony et al. 1979). In India, groundnut cultivation is mostly monoculture in major groundnut-growing states likes Andhra Pradesh, Gujarat, Tamil Nadu, Karnataka, Maharashtra, and Orissa. However, there is a great diversity in the cultural practices followed in different agroclimatic regions. Extensive cultivation in groundnut with introduction of new varieties and adoption of modern agro-techniques has brought a great deal of changes in the abundance of cosmopolitan and regional pests and their associated natural enemy complexes. In 1987, Pillai released a reduviid predator [*Platymeris laevicollis*] and pathogens [*Metarhizium anisopliae*] for the control of the scarabaeid *Oryctes rhinoceros* on coconuts in Kerala, India.

7.3.3.2 Groundnut

After a long time, Sahayaraj (1999b), Sahayaraj and Martin (2003), Sahayaraj (2002b) mass were reared *Rhynocoris marginatus* in the laboratory and released in the groundnut field and reported that greatly suppressed the population of *Spodoptera litura*, *Helicoverpa armigera*, and *Aproaerema modicella*. Sahayaraj (2002) made a comparison between a reduviid predator, *Rhynocoris marginatus* (Fab.), and water extracts of neem, *Pongamia*, *Vitex*, and *Calotropis* leaves at 3 % which were evaluated for their effectiveness against *Aproaerema modicella* and *Spodoptera litura* in groundnut (TMV-7 variety) field. *Rhynocoris marginatus* drastically reduced *Spodoptera litura* but not *Aproaerema modicella*. Among the four plant extracts integrated with the reduviid predator, the highest yield (1686 kg ha^{-1}) of observation was with neem. The net return per ha and cost-benefit ratio were also highest in groundnut + predator + neem treatment followed by groundnut + predator + *Pongamia*, groundnut + predator, groundnut + predator + *Vitex*, and groundnut + predator + *Calotropis*. There was no apparent effect of plant product treatments on *Rhynocoris marginatus* populations.

Considering the importance of integrated pest management (IPM), as a first step, the augmentative release of a reduviid predator *Rhynocoris kumarii* Ambrose and Livingstone and hot water extract of *Ipomoea carnea* Jacq. (family Convolvulaceae) and *Vitex negundo* Linn. (family Verbenaceae) were evaluated against selected groundnut pests at farmer's field. The results revealed that invariably endosulfan highly reduces all the pest populations. *Aphis craccivora* (Koch) (Hemiptera: Aphididae) population was highly reduced by the combination of both plant extracts in equal proportion (54.40 %) followed by *Ipomoea cornea* (48.86 %) and *Vitex. negundo* (37.85 %). Similar observations were also observed for *Helicoverpa armigera* Hübner (Lepidoptera: Noctuidae) and *Spodoptera litura* (Fab.) (Lepidoptera: Noctuidae). Both the plants have very limited impact on grasshoppers such as *Chrotogonus trachypterus* and *Atractomorpha crenulata*. *Rhynocoris kumarii* highly reduce the population of *Spodoptera litura* followed by *Aphis craccivora*, *Helicoverpa armigera*, *Chrotogonus trachypterus* Blanch., and *Atractomorpha crenulata* Fab. But the reduviid predator has the ability to suppress *Mylabris pustulata*.

Among all the treatments tested, groundnut production was higher in field-released *Rhynocoris kumarii* followed by endosulfan and the combination of both plant extracts in equal proportion sprayed on plot. In general, all the treatments invariably reduce the pest populations and increase the groundnut production, and hence they can be utilized and incorporated in the groundnut pests' managements (Sahayaraj and Ravi 2007a). There was no difference in the number of predatory arthropods, such as *Menochilus* sp., *Orius* sp., *Geocoris* sp., *Cantheconidia* sp., *Rhynocoris fuscipes*, *Menochotis* sp., and spiders, found in released and non released field plots. The percentage of good-quality cotton and yield of seed cotton were greater in released plots than in non released control plots.

7.3.3.3 Cotton

In India, during 2001, Claver and Ambrose evaluated the biocontrol potential of *Rhynocoris kumarii* Ambrose and Livingstone against *Helicoverpa armigera* (Hübner), *Spodoptera litura* (Fab.), and *Euproctis mollifera* Thunberg in cotton field cages. The adult *Rhynocoris kumarii* significantly suppressed *Helicoverpa armigera* larval population during the initial infestation, but the subsequent suppression of *Helicoverpa armigera* larvae was not significant. Though the adult and fifth instars of *Rhynocoris kumarii* minimized the infestation in *Spodoptera litura*- and *Euproctis mollifera*-infested cages, the suppression was not significant in *Helicoverpa armigera*-infested cages than in the control cages. However, *Rhynocoris kumarii* signific *Helicoverpa armigera* infested cages. Again in 2002, *Rhynocoris marginatus* was released in large-sized cotton field cages against three cotton pests, *Spodoptera litura*, *Mylabris pustulata*, and *Dysdercus cingulatus*. *Rhynocoris marginatus* highly reduced the infestation of *Spodoptera litura* (57.5 %), *Mylabris pustulata* (52.3 %), and *Dysdercus cingulatus* (45.8 %). The leaves, flower, and boll damages (32 %, 35 %, and 28 % by *Spodoptera litura*, *Mylabris pustulata*, and *Dysdercus cingulatus*, respectively) and seed cotton yield loss (1.4, 1.6, and 1.25 times in *Spodoptera litura*-, *Mylabris pustulata*-, and *Dysdercus cingulatus*-infested cages, respectively) were reduced by *Rhynocoris marginatus* compared with such fields without *Rhynocoris marginatus*. The field observations suggest this predator's pest suppression and damage reduction efficacy.

The present study was undertaken to find out the impact of the different prey-reared *Rhynocoris marginatus* separately on groundnut pest infestation, yield, and cost-benefit ratio in groundnut fields. During 2001, Claver and Ambrose took an attempt to augment *Rhynocoris kumarii* Ambrose and Livingstone in a cotton field to suppress *Dysdercus cingulatus* (Fabricius). The first release of adult stage of *Rhynocoris kumarii* did not affect the number of adult *Dysdercus cingulatus* in the released and control field plots. The second release of *Rhynocoris kumarii* in egg stage did not affect the number of first instar *Dysdercus cingulatus*. The third release of *Rhynocoris kumarii* in the third instar did not affect the number of second and third nymphal instars of *Dysdercus cingulatus*. However, after the fourth release of the fifth instar stage of *Rhynocoris kumarii*, there were significantly fewer fourth, fifth, and adult *Dysdercus cingulatus* in predator released plots compared to nonreleased (control) plots. *Dysdercus cingulatus* plant damage was less severe in predator released plots than in nonreleased control plots.

In 2003, Claver and his co-workers conducted a field experiment at the Agricultural College Farm, Killikulam, South India, with an assassin bug *Rhynocoris kumarii* Ambrose and Livingstone after mulching with sorghum trash and coconut leaflets and with shelter provisioning with pieces of clay pots and stones in a cotton field. The third and fourth stadium of *Rhynocoris kumarii* were released. There were fewer *Helicoverpa armigera* Hübner larvae in plots with mulched cotton trash than in control and other (mulched and shelter provisioned) plots. But mulching did not affect the number of adult *Mylabris pustulata* Thunberg. The flower and boll damage was significantly less in trash and leaflet mulch plots than other shelter was provisioned and control plots. The percentage of good-quality cotton was also greater than in control plots. The yield of seed cotton was also greater than in plots with trash mulches and coconut leaflet mulches than in control plots.

7.3.4 Australia

Grundy and Maelzer (2000a) evaluated biological control of *Pristhesancus plagipennis* in the pigeon pea field and reported the control of various pests in Australia. In Australia, Grundy and Maelzer (2000a, b, c) released third-instar nymphs of the Australian assassin bug, *Pristhesancus plagipennis* (Walker), into cotton and soybean plots at a number of release densities to test their biological control potential. *Helicoverpa* spp. and mirids, *Creontiades* spp., have been difficult to control biologically in cotton due to their unpredictable temporal abundance combined with a cropping environment often made hostile by frequent usage of broad-spectrum insecticides. They recorded that the number of *Helicoverpa* spp. larvae in cotton was reduced with the release of three or more *Pristhesancus plagipennis* nymphs per m crop row. But in soybean ecosystem, populations of green mirids, *Creontiades dilutus* (Stål), and looper caterpillars, *Chrysodeixis* spp., were reduced when *Pristhesancus plagipennis* were released at two nymphs per m crop row. The research suggests that *Pristhesancus plagipennis* has the capacity to be an effective biological control agent of soybean and cotton insect pests when released inundatively (Grundy and Maelzer 2000a, b, c). Later in 2002, third-instar nymphs of the Australian assassin bug, *Pristhesancus plagipennis* (Walker), were released into cotton plots at two release densities (0.51 and 1.38 nymphs per meter row) and two crop growth stages to test their biological control potential (Grundy and Maelzer 2002). Results showed that over 70 % of nymphs died within 2 weeks of release. Crop yields were significantly greater in the plots to which *Pristhesancus plagipennis* nymphs were released.

Carpenter and Greany (1998) were showed that medium-reared *Diapetimorpha introita* searched for and parasitized host pupae in field-cage experiments. These authors suggested that the ability to rear *Diapetimorpha introita* on an inexpensive artificial medium enhanced the possibility of mass-rearing parasitoid for use of inundative releases against *Spodoptera* spp. Host location and host acceptance by female parasitoids can be influenced by maternal factors such as age (Doutt 1959), physiological state, and previous experience (Morrison and King 1976). However, until the development of effective artificial rearing systems, laboratory-reared parasitoids completely lacking, both as developing immatures and as adults, were not available.

The crop field and screenhouse release of the reduviid predators that were an efficacious biological control agent to suppress the outbreak of various arthropod herbivores have been discussed in the previous sections. Due to performance in pest management, studies have been performed on aspects of biology, physiology, behavior and ecology of *Rhynocoris marginatus*. Moreover, the biological control population needs abundant ladybirds for inundative or inoculative release, which may influence the efficiency of pest management.

7.4 Bioefficacy of Reduviid Reared Artificially

The increasing commercial demands of mass biological control agents for agricultural production have intensified the need for effective artificial diets, not only in providing sufficient insect nutrition and rearing efficiency.

To evaluate the bioefficacy of the reduviid, the experiment was carried out in farmer's field from mid-December 2006 to May 2007. Groundnut (variety – TMV-7) was cultivated under well irrigation. The farmer was advised not to use pesticides or any other pest control practices during the experimental time. A total area of 4,050 m^2 (36 × 45 m) was chosen for the present study. It was divided into nine plots. Three treatments were performed in this study. They were:

T_1 – *Rhynocoris marginatus* reared with *Corcyra cephalonica*
T_2 – *Rhynocoris marginatus* reared with artificial diet
T_3 – Control (plot without predator)

In plots T_1 and T_2, *Rhynocoris marginatus* life stages except fifth instar (360 predators/subplot) were released (60 each) on the 40th day after seedling emergence (ASE). Totally 1,080 *Rhynocoris marginatus* were released during the study period. This release was done in the early morning hours (6:30 a.m. to 8:30 a.m.). One of the plots served as control where no predators were released. On the release day, the laboratory-reared 24 h-starved *Rhynocoris marginatus* were released into the field using a Camlin brush (2.0 cm) into the topmost five compound leaves.

7.4.1 Pest Sampling

The sampling of the pest was done by visual observation and expressed in number of pests/plant. For the infestation of the pests, the uppermost ten leaves were considered (Amin 1983), and 90 plants were reared randomly in each plot. The sampling was done 4 days before and after the release of the predators.

7.4.2 Cost-Benefit Ratio (CBR) Analysis

To estimate the cost-benefit ratio (CBR) and percent avoidable loss, on the harvest day, 90 plants were selected randomly from each plot, and one pod, two pods, and three pods in each plant were recorded. After the harvest, the production of groundnut from each plot was estimated and expressed in $Kgha^{-1}$. The CBR was also worked according to Kalyanasundaram et al. (1994). Percent avoidable losses were calculated using the formula of Krishnaiah (1980). Later, the effects of the different diets on developmental times and adult weights of *Rhynocoris marginatus* were analyzed using paired sample *t*-test ($P = 0.05$). Diet effects on reproductive parameters of *Rhynocoris marginatus* were also tested using one-way ANOVA and paired sample *t*-test. A student's *t*-test was applied to compare reproductive characters of *Rhynocoris marginatus* between *Corcyra cephalonica* and oligidic diets. Chi-square test was used to find out the significance for sex ratio (SPSS 13 version).

7.4.3 Pest Population

The arthropod-predator complex recorded in groundnut fields was highly diverse. It found that total pest populations (Table 7.1) suppress strengthened significantly when predator communities compared to the average predators species in groundnut field, and this effect was independent of prey diversity. Our experimental design included both artificial diet- and factitious prey-reared predators and maintained constant relative abundance of different species, in both experimental plots. A separate issue is whether any single predator species could achieve control equal to that observed by the diverse predators. For example, in our experiment, *Coccinella septempunctata* did not support the aphid population, and hence it's decreased when *Coccinella septempunctata* increased; it was proved by statistical analyses significance was observed. When such circumstances occur, conservation of the single most effective species could improve biological control as effectively as conserving the entire predator's assemblage (Straub and Snyder 2006).

As observed in *Coccinella septempunctata*, both artificial diet- and natural diet-reared *Rhynocoris marginatus* insects suppressed *Spodoptera litura* and *Helicoverpa armigera* populations. Similar observations were also observed by Sahayaraj (1999b, 2002b). Claver and Ambrose (2001) also reported that *Rhynocoris kumarii* suppressed the lepidopteron pest such as *Spodoptera litura* and *Helicoverpa armigera* in the cotton field. The field-cage experiment was also observed by Ambrose and Claver (1999b). It was also reported that *Rhynocoris kumarii* suppressed the lepidopteran pest population such as *Spodoptera litura* and *Helicoverpa armigera*. Whereas *Rhynocoris marginatus* provide strong suppression in experiment, diverse predatory effect on different pest communities exerted strong control on both experiments, suggestions that predator *Rhynocoris marginatus* biodiversity may provide the best result that effective lepidopteran and also other pest control results (Sahayaraj and Ravi 2007a).

Sahayaraj (2002a, b) reported that *Rhynocoris marginatus* mainly controlled *Spodoptera litura* than *Aproaerema modicella*. Present result revealed *Rhynocoris marginatus* mainly control jassids, *B. tabaci* (CC), *Mylabris indica*, *Myllocerus* spp. (CC), *N. viridula* (CC), and *Spodoptera litura* (CC) in summer and *P. ricini*, grasshoppers, *Mylabris indica*, *Spodoptera litura*, *Helicoverpa armigera*, and *Aphis*

Table 7.1 Impact of *Corcyra cephalonica*- and artificial diet (AD)-reared *Rhynocoris marginatus* on the incidence of groundnut pests (no./plant) in summer season

		DASE						
Pests	Treatment	36	43	50	57	64	71	Mean
Aphis craccivora	Control	88.9	36.1	32	22.7	27.3	21.4	38.08
	AD	96.5	96.0	66.2	44.1	28.00	21.8	55.13
	CC	86.6	54.2	32.4	31.3	28.70	18.5	41.95
Helicoverpa armigera	Control	1.0	0.4	0.0	0.0	0.7	0.3	0.40
	AD	0.7	4.3	1.3	0.4	0.3	0.0	1.16
	CC	1.9	0.9	0.4	0.7	0.0	0.0	0.65
Spodoptera litura	Control	0.1	0.4	3.0	0.0	1.7	0.0	0.86
	AD	3.3	2.6	1.5	1.0	0.9	0.0	1.55
	CC	0.9	0.0	2.4	0.0	1.6	0.0	0.81
Nezara viridula	Control	0.3	0.0	0.2	0.3	0.0	0.2	0.16
	AD	0.4	0.3	0.2	0.0	0.0	0.0	0.15
	CC	0.0	0.0	0.3	0.1	0.4	0.2	0.11
*Mylloceru*s sp.	Control	0.7	1.5	0.0	0.2	1.6	0.0	0.66
	AD	2.2	1.5	0.0	0.0	0.0	0.0	0.61
	CC	0.0	0.0	0.4	0.0	0.2	0.0	0.10
Mylabris indica	Control	3.3	1.3	2.6	2.7	4.2	2.2	2.71
	AD	4.6	2.4	3.7	3.2	3.5	1.7	3.18
	CC	2.2	1.8	1.2	1.3	1.9	1.6	1.51
Grasshoppers	Control	2.8	4.7	2.1	1.6	3.6	3.2	3.00
	AD	6.0	5.8	5.4	5.2	4.2	3.0	4.93
	CC	5.4	4.2	4.2	3.2	2.6	2.2	3.63
B. tabaci	Control	8.6	1.5	12	0.0	0.0	0.0	3.68
	AD	8.3	9.7	7.6	0.0	0.0	0.0	4.26
	CC	6.4	3.8	0.0	0.0	0.0	0.0	1.70
Jassid	Control	22.0	23.5	17.4	18.0	11.1	11.25	17.20
	AD	21.9	17.3	8.7	12.0	8.0	5.0	12.15
	CC	14.7	13.6	10.5	10.2	5.0	5.6	9.93

craccivora in kharif. This study indicated that *kharif* was unfavorable for the establishment (mean population was 2.61 and 2.91 for CC and artificial diet categories, respectively) and also control of many pests in summer. The composition of the natural enemy complex differed between maize fields, with certain predators favored over others, depending on the location of these fields. Despite this high variability in abundance and diversity of the predator complex, this finding suggested the role of *Rhynocoris marginatus*, *Rhynocoris fuscipes*, *Rhynocoris longifrons*, and *Coccinella septempunctata* in groundnut pest management.

The field release of *Pristhesancus plagipennis* (Grundy and Maelzer 2000a, b, c) and *P. laevicollis* (Antony et al. 1979) was successful in reducing various pests in their released fields. Sahayaraj (2002a, b) and Sahayaraj and Ravi (2007a, b) integrated *Rhynocoris marginatus* along with some botanicals in groundnut field, and he observed the drastic reduction of pests such as *Spodoptera litura* and *Helicoverpa armigera* and achieved a high yield of groundnut. *Rhynocoris kumarii* suppressed various pests in the groundnut field as reported in the laboratory studies (Ambrose 1996; Claver 1998). But it did not reduce the population of *Mylabris pustulata*. But Ambrose (1999) reported that it feeds on *Mylabris pustulata* in the laboratory. Present study shows that *Rhynocoris marginatus* during kharif reduced *Mylabris indica* more than 75 %.

During the study period, five pests such as *Aphis craccivora* Koch, *Helicoverpa armigera*, *Spodoptera litura*, *Mylabris indica*, and grasshoppers (*Atractomorpha crenulata* and *Chrotogonus trachypterus*) were recorded both in summer (Table 7.1) and *kharif* 2006 and 2007. *Nezara viridula*, *Myllocerus* sp., *Bemisia tabaci*, and jassids were present only during summer, whereas *A. annulipes* and *Pericallia ricini* were present in *kharif* (Table 7.2). During this period, the most abundant pest was *Aphis craccivora* followed by jassids and grasshoppers. *Aphis craccivora* population was also prevalent in *kharif*. Correlation analyses revealed that during this period, the most abundant pest was *Aphis craccivora* followed by jassids and grasshoppers. Four of them were sucking pests and others were defoliators. In the field-released AD-reared *Rhynocoris marginatus*, *Aphis craccivora* was recorded as the predominant pest followed by jassids and grasshoppers. The incidence of the entire pest decreased significantly when the groundnut grew older.

Two grasshoppers ($t = 0.22$, $df = 9$, $P \geq 0.82$) were recorded in both seasons. Among the treatments, grasshopper population was higher and lower in OD and control categories during summer and *kharif* season, respectively. Similar trend was also observed for *Aphis crassivora*, *Helicoverpa armigera* (1.16, $t = 1.59$, $df = 9$, $p \geq 0.145$), *Spodoptera litura* ($t = 1.500$, $df = 9$, $p \geq 0.167$), and *Mylabris indica* (Tables 7.1 and 7.2). But jassid, *Nezara viridula*, and *Myllocerus* sp. population was high in control followed by OD and *Corcyra cephalonica* categories. *Pericallia ricini* (t-0.0276, $p \geq 0.978$) population was maximum and minimum in control and OD categories (2.78/plant).

7.4.4 Natural Enemies

Among all predators recorded, *Rhynocoris marginatus* was the most abundant to found the gradual increases when groundnut plant was maturing. Though spider's population was found to be of lower densities, it was maintained throughout the groundnut vegetative cycle. When release rate of *Rhynocoris marginatus* increased, the ratio of the number of prey per natural enemy decreases. From the results, it was very clear that *Rhynocoris marginatus* greatly suppressed the population of *Spodoptera litura*, *Aphis craccivora*, *Helicoverpa armigera*, and *Pericallia ricini*. Spiders and *Coccinella septempunctata* were the most abundant predators in groundnut during summer (Table 7.3). But *Rhynocoris marginatus* was dominated in *kharif* (Table 7.4). Among the two-diet regime, artificial diet-reared *Rhynocoris marginatus* slightly and highly reduce the pest populations.

After three generations, reduviids reared on the meat-based diet were as active at preying on the target preys, *Corcyra cephalonica*, *Dysdercus cingulatus*, and

Table 7.2 Impact of factitious (T_1) and artificial diet (T_3)-reared *Rhynocoris marginatus* on the incidence of groundnut pests (no./plant) in *kharif*

Pests	Treatment	DASE 36	43	50	57	64	71	Mean
Aphis craccivora	Control	27.4	27.4	6.0	4.9	9.5	5.0	13.36
	Artificial diet	21.6	17.6	8.1	1.2	0.6	0.0	8.18
	CC	2.4	6.0	3.0	1.6	0.0	0.0	2.16
Helicoverpa armigera	Control	3.2	0.6	0.3	0.7	0.4	0.6	5.2
	Artificial diet	1.2	1.4	0.5	0.4	0.0	0.0	0.58
	CC	3.6	1.4	1.5	0.9	0.0	0.0	1.23
Spodoptera litura	Control	4.0	6.5	4.2	0.1	0.5	0.0	2.55
	Artificial diet	2.6	1.9	0.4	0.2	0.0	0.0	0.85
	CC	2.2	0.4	0.6	0.4	0.0	0.0	0.6
Ants	Control	4.5	4.1	2.8	1.2	1.6	0.5	14.7
	Artificial diet	4.0	3.8	4.1	1.0	0.0	0.0	12.9
	CC	4.3	2.6	2.9	1.7	2.0	0.0	13.5
Anisolabis annulipes	Control	0.6	0.4	0.3	0.2	0.0	0.0	0.25
	Artificial diet	1.6	0.3	0.2	0.3	0.4	0.5	0.40
	CC	0.8	0.4	0.2	0.3	0.0	0.0	0.28
Mylabris indica	Control	5.6	0.7	3.3	3.8	1.4	1.1	15.9
	Artificial diet	2.1	0.2	1.5	0.9	0.2	0.0	0.81
	CC	1.1	0.5	0.4	0.8	0.4	0.0	3.2
Grasshoppers	Control	4.8	3.6	3.1	4.6	2.9	3.8	3.8
	Artificial diet	3.3	1.0	0.7	0.5	0.3	0.0	0.96
	CC	2.8	2.0	0.9	0.9	1.0	0.0	1.26
Pericallia ricini	Control	1.1	8.0	10.4	4.2	1.1	0.8	4.26
	Artificial diet	9.6	3.6	2.5	1.0	0.0	0.0	2.78
	CC	10.4	4.2	1.5	1.0	0.8	0.0	2.98

S. litura, than those tested, in oligidic diet. This indicated that feeding of *Rhynocoris marginatus* with oligidic diet did not alter their capacity to recognize and feed on their natural and factitious preys. Other artificially reared predators recognized and accepted their target preys, as shown by Hagler and Cohen (1991) and Cohen (2000) for *Geocoris punctipes*, by Chocorosqui and De Clercq (1999) and De Clercq and Degheele (1993) for *Podisus maculiventris* and *Podisus sagitta*, by Castane et al. (2002) for *Dicyphus tamaninii*, and by Castane and Zapata (2005) for *Macrolophus caliginosus*.

7.4.5 Production Data

Among all the treatments, number of two pods was more than three pods in a groundnut plant. For instance, in T_1 and T_4 categories, the number of pods per plant was 24.17 and 71.79, respectively. Three pods were not observed during summer

Table 7.3 Impact of *C. cephalonica*- and artificial diet (T_3)-reared *Rhynocoris marginatus* on the incidence of groundnut predator (no./plant) in summer season

Predators	Treatments	DASE						Mean
		36	43	50	57	64	71	
Rhynocoris marginatus	Control	0.0	0.0	0.0	0.1	0.2	0.2	0.08
	Artificial diet	0.2	0.3	0.4	0.7	1.1	2.1	0.80
	CC	0.0	0.2	0.3	0.6	0.9	1.5	0.58
Rhynocoris fuscipes	Control	0.1	0	0	0.1	0	0.1	0.05
	Artificial diet	0.0	0.1	0.0	0.5	0.5	1.1	0.28
	CC	0.0	0.3	0.6	0.1	0.0	0.5	0.25
Rhynocoris longifrons	Control	0.0	0.0	0.0	0.1	0.0	0.0	0.01
	Artificial diet	0.0	0.0	0.0	0.1	0.1	0.1	0.05
	CC	0.0	0.0	0.0	0.0	0.0	0.0	0.00
Spiders	Control	0.3	0.2	0.5	0.1	0.3	0.0	0.23
	Artificial diet	0.1	0.2	2.4	4.6	3.3	4.6	2.53
	CC	1.1	1.2	1.4	1.4	2.4	0.7	1.20
Dragonfly	Control	0.2	1.0	0.1	0.2	1.0	0.0	0.08
	Artificial diet	0.4	0.0	0	1.1	0.4	0.2	0.35
	CC	0.7	0.4	0.2	0.4	0.6	0.5	0.46
Mantis religiosa	Control	0.1	0.2	0.0	0	0.0	0.1	0.06
	AD	0.3	0.0	0.0	0.4	0.1	0.1	0.15
	CC	0.0	0.5	0.5	0.4	0.0	0.4	0.30
Coccinella septempunctata	Control	0.3	0.2	0.3	0.1	0.0	0.2	0.18
	AD	0.8	2.1	1.4	1.7	1.7	0.7	1.16
	CC	1.0	1.0	0.5	0.9	1.1	0.5	0.83

2007. The oligidic diet-reared predator plots were two pods/plant higher (Table 7.5). During summer 2006, groundnut production was maximum in the oligidic diet-reared predator plot (1,224 kg/h) followed by T_1 (936.00 kg/h) and minimum in the control plot (728.00 kg/h). Same trend was also observed in *kharif* 2007. Similarly the cost-benefit ratio was high in the field-released T_3 (artificial diet) predator (1:2) followed by T_1 (CC)-reared predator (1:1.8). Similar results were observed in *kharif* 2007 too. During 2006, the percent avoidable loss (PAL) was maximum (23.33 %) in field-released artificial diet-reared predator, followed by T_1-reared predator released (14.81 %). So *Rhynocoris marginatus* could be used as a biological control agent against groundnut pests (Table 7.5).

Cost benefit ratio at predator released in the field to enhanced the production of groundnut (Fig. 7.1) and cost benefit ratio it. These agreements are findings of Sahayaraj (1999b, 2002b) and Sahayaraj and Martin (2003) (Sahayaraj and Ravi 2007a) on *Rhynocoris marginatus* released in groundnut field. So *Rhynocoris marginatus* (reared with artificial diet and natural diet) could be used as a biological control agent against groundnut pests. This work provides the basis for an exploration of opportunities to conserve natural enemies and increase biological control potential (Barbosa 1998; Landis et al. 2000). Many of the natural enemies identified

Table 7.4 Impact of factitious host (T_1)- and artificial diet (T_3)-reared *Rhynocoris marginatus* on the incidence of groundnut predator (no./plant) in *kharif*

		DASE						
Predators	Treatments	36	43	50	57	64	71	Mean
Rhynocoris marginatus	Control	1.0	0	1.4	0.9	0.3	0.3	0.65
	AD	0.2	1.9	2.4	3.6	4.1	5.3	2.91
	CC	1.1	2.1	2.7	3.0	2.8	4.0	2.61
Rhynocoris fuscipes	Control	0.4	0.0	0.2	0.3	0.3	0.3	0.25
	AD	0.0	0.4	0.0	0.0	0.5	1.1	0.33
	CC	00	0.0	0.1	0.1	0.3	0.5	0.16
Rhynocoris longifrons	Control	0.9	0.0	0.7	0.4	0.3	0.1	0.40
	AD	0.0	1.2	0.2	0.1	0.1	0.1	0.28
	CC	0.0	0.0	0.2	0.0	0.4	0.6	0.20
Spider	Control	0.5	0.5	0.4	0.7	1.5	1.4	0.66
	AD	1.0	0.4	0.7	0.5	2.2	1.8	0.93
	CC	0.3	0.4	0.8	0.7	0.9	0.5	0.93
Dragonfly	Control	0.7	0.2	0.6	0.2	1.7	0.0	0.56
	AD	0.4	0.3	0.2	0.2	0.4	0.2	0.28
	CC	0.3	0.3	0.7	0.3	0.5	0.6	0.45
Mantis religiosa	Control	0.2	0.1	0.1	0.1	0.0	0.3	0.13
	AD	0.0	0.3	0.1	0.0	0.1	0.1	0.10
	CC	0.1	0.0	0.4	0.1	0.4	0.5	0.25
Coccinella septempunctata	Control	0.6	0.0	0.5	0.5	0.3	0.5	0.40
	AD	0.9	0.7	0.7	0.5	1.7	0.7	0.86
	CC	0.7	0.2	0.9	0.7	1.0	0.6	0.68

Table 7.5 *C. cephalonica* (T_1)- and artificial diet (T_3)-reared *Rhynocoris marginatus* and control on the groundnut production (2007)

Treatments	Total no. of pods/ plant	One pod/ plant	Two pods/ plant	Weight of total pods/ plant
T_3	58.29	4.13	54.16	78.43
T_1	45.66	8.43	37.23	66.26
Control	47.7	6.1	41.6	69.83

from this study having potential for pest biological control. However, their conservation practices targeting those await and utilize in subsistence groundnut production systems.

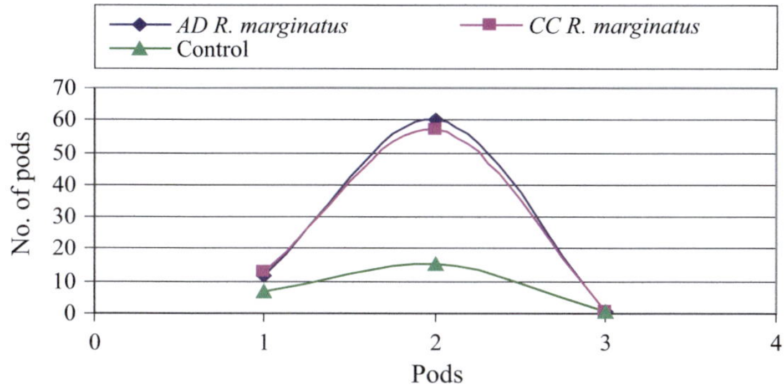

Fig. 7.1 Groundnut production in natural diet-reared and artificial diet-reared *Rhynocoris marginatus* and control

7.5 Conclusions

Among all predators recorded, *Rhynocoris marginatus* was the most abundant in the groundnut field. During the study period, five pests such as *Aphis crassivora* Koch, *Helicoverpa armigera*, *Spodoptera litura*, *Mylabris indica*, and grasshoppers were recorded both in summer and *kharif* 2006 and 2007. During this period, the most abundant pest was *Aphis crassivora* followed by jassids and grasshoppers. From this result, it was very clear that *Rhynocoris marginatus* greatly suppressed the population of *Spodoptera litura*, *Aphis craccivora*, *Helicoverpa armigera*, and *Pericallia ricini*. Among the two-diet regime, *Rhynocoris marginatus* reared with artificial diet highly reduce the pest populations. Both in summer 2006 and *kharif* 2007, production of the groundnut was high in the artificial diet-reared predator. Similarly the cost-benefit ratio was maximum in the field-released artificial diet-reared predator (1:2) followed by T_1 (CC)-reared predator (1:1.8).

In the present study, *Rhynocoris marginatus* reared with artificial diet greatly suppressed the population of *Spodoptera litura*, *Aphis craccivora*, *Helicoverpa armigera*, and *Pericallia ricini*. Biological control potential of *Rhynocoris marginatus* was not similar in both seasons. Moreover, *Coccinella septempunctata* and spiders were also established during this season. However, in summer *Rhynocoris fuscipes*, spiders, dragonfly, *M. religiosa*, and *Coccinella septempunctata* were established.

Among the two-diet regime, *Rhynocoris marginatus* slightly and highly reduce the pest populations. This study is discussed for most augmentative biological control agents; there is a rearing of *Rhynocoris marginatus* on artificial diet that produces effective control of a pest species. This study was especially true when predators were used in biological control. It is clear that artificial diet-reared predators should be carefully considered before implementing augmentative biological control efforts with natural diet-reared insects.

The ultimate success of augmentative biological control may depend on releases of biological control agents that maximize establishment, are released in synchrony with the host, and can be integrated into integrated pest management program. Thus, determination of OD-reared *Rhynocoris marginatus* that maximize the effectiveness of natural enemies can increase the effectiveness of augmentation biological control and increase potential economic benefits.

7.6 Future Area

- Whether increased rate of release does improve control or releasing the fewer of natural enemies. It will be efficiently and economically beneficial augmentative biological control.
- Artificial diet can be useful for augmentative field release of predatory insects like reduviids or not.
- Whether reduviid can be used alone or in combination with other IPM components.

References

Ables JR (1978) Feeding behavior of an assassin bug, *Zelus renardii*. Ann Entomol Soc Am 71(4):476–478

Ambrose DP (1996) Assassin bugs (Insecta: Heteroptera: Reduviidae) in biocontrol: success and strategies, a review. In: Ambrose DP (ed) Biological and cultural control of insect pests an Indian scenario. Adeline Publishers, Tirunelveli, pp 262–284, pp–362

Ambrose DP (1999) Assassin bugs. Science Publishers, Enfield

Ambrose DP, Claver MA (1999) b. Suppression of cotton leaf worm *Spodoptera litura*, flower beetle *Mylabris pustulata* and red cotton bug *Dysdercus cingulatus* by *Rhynocoris marginatus* (Fabr.) (Heteroptera: Reduviidae) in cotton field cages. J Appl Entomol 123:225–229

Amin PW (1983) Major field and storage insect pests of groundnut in India and their control. Occasional paper 1/83, Groundnut improvement programme. ICRISAT, Patancheru

Antony M, Daniel J, Kurian C, Pillai GB (1979) Attempts in introduction and colonization of the exotic reduviid predator *Platymeris laevicollis* Distant for the biological suppression of the coconut rhinocerous beetle, *Oryctes rhinoceros*. Proc Plant Crops Symp 2:445–454

Aquilino KM, Cardinale BJ, Ives AR (2005) Reciprocal effects of host plant and natural enemy diversity on herbivore suppression: an empirical study of a model tritrophic system. Oikos 108:275–282

Barbosa PA (ed) (1998) Conservation biological control. Academic Press, San Diego

Bengtsson J, Ahnstrom J, Weibull AC (2005) The effects of organic agriculture on biodiversity and abundance: a meta-analysis. J Appl Ecol 42:261–269

Biswas GC, Begum S, Mian MY (2000) Leaf infestation and yield loss caused by jassid and thrips in groundnut. J Asiat Soc Bangladesh Sci 26(2):253–258

Blatchley WS (1926) Heteroptera, or true bugs of eastern North America; with special reference to the faunas of Indiana and Florida. Nature Publishing Company, Indianapolis

Byrnes J, Stachowicz JJ, Hultgren KM, Hughes AR, Olyarnik SV, Thornbert CS (2005) Predator diversity strengthens trophic cascades in kelp forests by modifying herbivore behaviour. Ecol Lett 9:61–71

Cardinale BJ, Harvey CT, Gross K, Ives AR (2003) Biodiversity and biocontrol: emergent impacts of a multiple-enemy assemblage on pest suppression and crop yield in an agroecosystem. Ecol Lett 6:857–865

Carpenter JE, Greany PD (1998) Comparative development and performance of artificially reared versus host-reared *Diapetimorpha introita* (Cresson) (Hymenoptera: Ichneumonidae) wasps. Biol Control 12:137–142

Castane C, Zapata R (2005) Rearing the predatory bug *Macrolophus caliginosus* on a meat-based diet. Biol Control 34:66–72

Castane C, Iriarte J, Lucas E (2002) Comparison of prey consumption by *Dicyphus tamaninii* reared conventionally and on a meat based diet. BioControl 47:657–666

Chocorosqui VR, De Clercq P (1999) Developmental and predatory performance of *Podisus maculiventris* (Say) (Heteroptera: Pentatomidae) reared on a meat-based artificial diet. Meded Faculteit Landbouwwetenschappen Rijksuniversiteit Gent 64:229–234

Claver MA (1998) Biological control potential evaluation of the predator, *Rhynocoris kumarii* Ambrose and Livingstone (Hemiptera: Reduviidae) against three selected cotton insect pests. Ph. D. Thesis. M. S. University, Tirunelveli, India

Claver MA, Ambrose DP (2001) Influence of hunger level and prey density on the searching behaviour of *Rhynocoris kumarii* Ambrose and Livingstone (Hemiptera: Reduviidae), a potential predator of *Spodoptera litura* Fabricius (Lepidoptera: Noctuidae). J Entomol Res 25(4):309–313

Cohen AC (2000) Feeding fitness and quality of domesticated and feral predators: effects of long-term rearing on artificial diet. Biol Control 13:49–54

De Clercq P, Degheele D (1993) Quality assessment of the predatory bugs *Podisus maculiventris* (Say) and *Podisus sagitta* (Fab.) (Heteroptera: Pentatomidae) after prolonged rearing on a meat—based artificial diet. Biocontrol Sci Tech 3(2):133–139

Doutt RL (1959) The biology of parasitic Hymenoptera. Annu Rev Entomol 4:161–182

El- Sebaey IIA, Abd El-Wahab HA (2003) Suppression of *Bemisia tabaci* (Genn), *Aphis gossypii* Glov. and *Spodoptera littoralis* (Bosid) by *Coranus africana* El-Sebaey (Het.: Reduviidae) in tomato field. Bull Fac Sci Cairo Univ 54(1):141–150

El- Sebaey IIA, Abd El-Wahab HA (2007) Evaluation the role of the assassin bug, *Coranus africana* El-Sebaey (Heteroptera: Reduviidae) in the suppression of different infestation levels of *Aphis gossypii* Glov. in Cucumber and squash fields. Egypt J Agric Res 85(2):489–496

El- Sebaey IIA, Abd El-Wahab HA, Ibrahim SA (2002) Suppression of white fly Bemisia tabaci (Genn.) with Augmentative release of assassin bug, Coranus africana El-Sebaey (Het; Reduviidae) in cucumber green house. J Unio Arab Biol Cairo 17((A):197–205

El- Sebaey IIA, Abd El-Wahab HA, Ibrahim SA (2004) Evaluation of the role of the predatory bug *Coranus africana* El-Sebaey (Heteroptera: Reduviidae) in the suppression of different infestation levels of the white fly *Bemisia tabaci* in tomato field. Al- Azhar J Agric Res 39:31–44

El-Sebaey IIA, El-Wahab HAA (2011) Effect of different stages of reduviid predator *Corans africana* El-sebaey hemiptera: heteroptera, on the population of *Bemisia tabaci* genn in tomato field at Qalubia and Bani-swif governorates. Egypt J Agric Res 89(2):413–430

Grundy P, Maelzer D (2000a) Assessment of *Pristhesancus plagipennis* (Walker) (Hemiptera: Reduviidae) as an augmented biological control in cotton and soybean crops. Aust J Entomol 39:305–309. doi:10.1046/j.1440-6055.2000.00182.x

Grundy P, Maelzer D (2000b) Predation by the assassin bug *Pristhesancus plagipennis* (Walker) (Hemiptera: Reduviidae) of *Helicoverpa armigera* (Hubner) (Lepidoptera : Noctuidae) and *Nezara viridula* (L.) (Hemiptera : Pentatomidae) in the laboratory. Aust J Entomol 39:280–282

Grundy P, Maelzer D (2000c) Assessment of *Pristhesancus plagipennis* (Walker) (Hemiptera: Reduviidae) as an augmented biological control in cotton and soybean crops. Aust J Entomol 39:305–309

Grundy PR, Maelzer DA (2002) Augmentation of the assassin bug *Pristhesancus plagipennis* (Walker) (Hemiptera: Reduviidae) as a biological control agent for *Helicoverpa* spp. in cotton. Aust J Entomol 41(2):192–196

Hagler JR, Cohen AC (1991) Prey selection by in vitro-and field-reared *Geocoris punctipes*. Entomol Exp Appl 59(3):201–205

Hart ER (1986) Genus *Zelus Fabricius* in the United States, Canada, and Northern Mexico (Hemiptera: Reduviidae). Ann Entomol Soc Am 79(3):535–548

Hole DG, Perkins AJ, Wilson JD, Alexanderd IH, Gricee PV, Evans AD (2005) Does organic farming benefit biodiversity? Biol Conserv 122:113–130

Jayanthi M, Singh KM, Singh RN (2000) Economics of insecticidal schedule against groundnut pests. Indian J Entomol 62(1):5–10

Kalyanasundaram M, Dhandapani N, Swamiappan M, Sundarababu PC, Jeyaraj S (1994) A study on the management of some pests of groundnut (*Arachis hypogaea* L.) with biocontrol agents. J Biol Control 8(1):1–4

Krishnaiah K (1980) Assessment of crop loses due to pests and diseases (Govindu HC, ed.). UAS Technical Series No. 33, pp 259–267

Landis DA, Wratten SD, Gurr GM (2000) Habitat management to conserve natural enemies of arthropod pests in agriculture. Annu Rev Entomol 45:175–201

Mead FW (1999) Wheel bug, *Arilus cristatus* (Linnaeus) (Insecta: Hemiptera: Reduviidae). University of Florida Cooperative Extension Service, Institute of Food and Agricultural Sciences, EDIS

Morrison RK, King EG (1976) Mass production of natural enemies. In: Ridgway RL, Vinson SB (eds) Biological control by augmentation of natural enemies. Plenum Press, New York, pp 183–217

Napompcth B, Winotai A, Sommartya P (1989) Utilization of natural enemies for biological control of the leucaena psyllid in Thailand. Leucaena Psyllid: Probl Manag 175–180

Panchabhavi KS, Nethradhaniraj CR (1987) Yield of groundnut as affected by varying larval density of *Spodoptera litura* Fabricius (Lepidoptera : Noctuidae). Indian J Agric Sci 57(7):525–527

Peter C, David BV (1998) Residual toxicity of some insecticides on the first and third instar larvae of *Spodoptera litura* F. (Lepidoptera : Noctuidae). Trop Pest Manag 34(1):24–26

Peveling R, Attignon S, Langewald J, Owambama Z (1999) An assessment of the impact of biological and chemical control agents on groundnut dwelling arthropods in Nigar, based on presence and absence sampling. Crop Prod 18:323–339

Ramaraju K, Rajamanicakam K, Sridharan C (1998) Efficacy of insecticides and botanicals against pests of groundnut. Pestology 22(8):27–30

Richman DB, Hemenway RC, Whitcomb WH (1980) Field cage evaluation of predators of the soybean looper, *Pseudoplusia includens* (Lepidoptera: Noctuidae). Environ Entomol 9(3):315–317

Sahayaraj K (1999a) Effect of prey and their age on the feeding preference of *Rhynocoris marginatus* (Fab.). Int Arachis Newsl 19:39–41

Sahayaraj K (1999b) Field evaluation of *Rhynocoris marginatus* (Fab.) against two groundnut defoliators. Int Arachis Newsl 19:41–42

Sahayaraj K (2002a) Small-scale laboratory rearing of reduviid predator *Rhinocoris marginatus* (Fab.) (Hemiptera : Reduviidae) on *Corcyra cephalonica* Stainton larvae by larval card method. J Central Eur Agric 3(2):137–148

Sahayaraj K (2002b) Field bioefficacy of a reduviid predator *Rhynocoris marginatus* and plant products against *Aproaerema modicella* and *Spodoptera liltura* of groundnut. Indian J Entomol 64(3):292–300

Sahayaraj K, Martin P (2003) Assessment of *Rhynocoris marginatus* (Fab.) (Hemiptera : Reduviidae) as augmented control in groundnut pests. J Cent Eur Agric 4(2):103–110

Sahayaraj K, Ravi C (2007a) Evaluation of reduviid predators and plant products against chosen groundnut pests. Arch Phytopathol Plant Protect 40(4):281–290

Sahayaraj K, Ravi C (2007b) Small-scale mass production strategy for a reduviid predator *Rhynocoris longifrons* Stal (Heteroptera: Reduviidae). In: Guptha VK, Verma AK (eds)

Perspective in animal ecology and reproduction, vol 4. Daya Publishing House, New Delhi, pp 53–81. ISBN 9788170354598

Schaefer CW (1988) Reduviidae (Hemiptera : Heteroptera) as agents of biological control. In: Anantha subramanian KS, Venkatesan P, Sivaraman (eds) Bicovas, vol 1. Loyola College, Madras, pp 27–33

Snyder GB, Finke DL, Snyder WE (2008) Predator biodiversity strengthens aphid suppression across single- and multiple-species prey communities. Biol Control 44:52–60

Speranza S, Melo MC, Luna MG, Virla EG (2014) First record of *Zelus obscuridorsis* (Hemiptera: Reduviidae) as a predator of the South American tomato Leafminer, *Tuta absoluta* (Lepidoptera: Gelechiidae). Fla Entomol 97(1):295–297

Sridhar V, Mahto Y (2000) Pest and natural enemy complex of groundnut (*Arachis hypogaea* L.) CV. ICGS – 1 in the agroecosystem of Delhi. Indian J Entomol 62(4):335–340

Straub CS, Snyder WE (2006) Species identity dominates the relationship between predator biodiversity and herbivore suppression. Ecology 87(2):277–282

Usman A, Ofori K, Danquah EY, Offei SK, Ado SG (2015) Genetic analysis of groundnut rosette virus disease in groundnut (Archis hypogaea L.). Afr J Plant Sci 9(3):115–123

Van Driesche RG, Bellows TS (1996) Biological control. Chapman and Hall, New York

Wightman JA, Rao GVR (1993) A groundnut insect identification hand book for India, Information Bulletin No. 39. ICRISAT, Patancheru

Wilby A, Villareal SC, Lan LP, Heong KL, Thomas MB (2005) Functional benefits of predator species diversity depend on prey identity. Ecol Entomol 30:497–501

Wygodzinsky P (1949) Elenco sistematico de los reduviiformes americanos. Tucumán.

General Recommendations

1. Conduct trial and error experiments with modifications of artificial diet ingredients to determine the best diet which decreased the nymphal developmental period and nymphal and adult mortality and increased fecundity and hatchability, group feeding, etc.
2. How long the proportion of reduviid predator populations in an ecosystem will be necessary to be considered in future work?
3. Whether the presence of specific prey, for example, lepidopteran caterpillars, attracts reduviid predator into cotton crop (or any other crops where reduviid population is very high) will be undertaken for future studies.
4. More meaningfully elucidate the bioefficacy of reduviids in relation to daytime or seasons or crops or prey which will be studied.
5. It is essential to find out the phenomena of interlude predation (IGR) in reduviid predator at laboratory and various crop fields.
6. Single-plex PCR or multiplex PCR assay will be conducted to amplify one prey or several prey types, respectively, and to track out the preference of reduviids against economically important pests.

K. Sahayaraj, R. Balasubramanian, *Artificial Rearing of Reduviid Predators for Pest Management*, DOI 10.1007/978-981-10-2522-8

Species Index

K. Sahayaraj, R. Balasubramanian, *Artificial Rearing of Reduviid Predators for Pest Management*, DOI 10.1007/978-981-10-2522-8

Subject Index

K. Sahayaraj, R. Balasubramanian, *Artificial Rearing of Reduviid Predators for Pest Management*, DOI 10.1007/978-981-10-2522-8

MIX
Papier aus verantwortungsvollen Quellen
Paper from responsible sources
FSC® C105338

If you have any concerns about our products,
you can contact us on
ProductSafety@springernature.com

In case Publisher is established outside the EU,
the EU authorized representative is:
Springer Nature Customer Service Center GmbH
Europaplatz 3, 69115 Heidelberg, Germany

Printed by Libri Plureos GmbH
in Hamburg, Germany